Gene Regulation by Steroid Hormones IV

# Gene Regulation by Steroid Hormones IV

Edited by
A.K. Roy and J.H. Clark

With 75 figures

Springer-Verlag
New York Berlin Heidelberg
London Paris Tokyo

ARUN K. ROY
Department of Obstetrics and Gynecology, University of Texas Health Science Center, San Antonio, Texas 78284, USA.

JAMES H. CLARK
Department of Cell Biology, Baylor College of Medicine, Houston, Texas 77030, USA.

*Cover Picture:* Electromicrograph of a hybrid molecule between ovalbumin gene and ovalbumin mRNA. The intervening sequences are shown as loops. Courtesy of Dr. Eugene Lai, Baylor College of Medicine.

Library of Congress Cataloging-in-Publication Data
Gene regulation by steroid hormones IV / edited by A.K. Roy and J.H. Clark.
 p. cm.
 Proceedings of the fourth conference on hormones held at Meadow Brook in the fall of 1988.
 Includes index.
 ISBN 0-387-96999-3 (alk. paper) : $80.00 (est.)
 1. Steroid hormones—Physiological effect—Congresses. 2. Genetic
regulation—Congresses. 3. Steroid hormones—Receptors—Congresses.
I. Roy, A.K. (Arun K.) II. Clark, James H. (James Henry), 1932–
 [DNLM: 1. Gene Expression Regulation—congresses. 2. Receptors,
Endogenous Substances—physiology—congresses. 3. Steroids–
–physiology—congresses. WK 150 G3261 1988]
QP572.S7G463 1989
574.87'322—dc20
DNLM/DLC
for Library of Congress             89-6271

Printed on acid-free paper.

Text provided by the editors in camera-ready form.
Printed and bound by Edwards Brothers, Inc., Ann Arbor, Michigan.
Printed in the United States of America.

9 8 7 6 5 4 3 2 1

ISBN 0-387-96999-3 Springer-Verlag New York Berlin Heidelberg
ISBN 3-540-96999-3 Springer-Verlag Berlin Heidelberg New York

# Preface

The first Meadowbrook Symposium was held in 1978 and during the intervening ten years our knowledge concerning how steroid hormones function at the level of gene expression has advanced by leaps and bounds. In this volume, which summarizes our fourth meeting, these advances are very evident. What seemed like science fiction ten years ago has become commonplace science. Who would have imagined that we could synthesize a nucleotide sequence that binds a specific steroid receptor and acts as a controlling element for gene expression? No one; but as is evident from the results reported in several chapters, this technique is yielding a wealth of information. Using these and other techniques it has become apparent that gene transcription is controlled by interactions between transacting factors and DNA recognition sequences (response elements). These transacting factors appear to be members of a large gene family that includes steroid hormone receptors, transcription factors, protooncogenes and homeobox proteins. Thus a great deal has been learned, but as usual, questions remain. Many of these questions are posed by the findings and observations found in several chapters in this volume. Non-hormone binding forms of steroid receptors and their relevance to receptor down regulation, recycling and biological response remain a mystery. The quantitative relationship between receptor binding and biological response still presents agonizing problems. These and many other intriguing questions are discussed in this volume and set the stage for what should be a most rewarding time in endocrinology.

Winter 1989

ARUN K. ROY
JAMES H. CLARK

# Contents

# Contributors

W. ANKENBAUER, Institute of Cell and Tumor Biology, German Cancer
    Research Center, 6900 Heidelberg 1, Federal Republic of Germany

D. BELLINGHAM, Department of Biochemistry, University of North Carolina,
    Chapel Hill, NC 27599, USA

M. BODNER, Department of Pharmacology M-036, University of California-San
    Diego, School of Medicine, LaJolla, CA 92093, USA

S. BOURGEOIS, Department of Regulatory Biology, The Salk Institute, 10010
    North Torrey Pines Road, LaJolla, CA 92037, USA

K. BURNSTEIN, Cancer Research Center, University of North Carolina, Chapel
    Hill, NC 27599, USA

J.-L. CASTRILLO, Department of Pharmacology M-036, University of California-
    San Diego, School of Medicine, LaJolla, CA 92093, USA

B. CHATTERJEE, University of Texas Health Science Center, Department of
    Cellular and Structural Biology, San Antonio, TX, USA

C. CHIAPPETTA, Department of Pharmacology, The University of Texas Medical
    School, Houston, TX 77225, USA

J.A. CIDLOWSKI, Department of Physiology and Biochemistry, University of
    North Carolina, 460 Medical School Research Building, Chapel Hill,
    NC 27514, USA

J. CLARK, Department of Cell Biology, Baylor College of Medicine, Houston,
    TX 77030, USA

O. CONNEELY, T. CROWE, Department of Cell Biology, Baylor College of
    Medicine, 1 Baylor Plaza, Houston, TX 77030, USA

W.F. DEMYAN, University of Texas Health Science Center, Department of
    Cellular and Structural Biology, San Antonio, TX, USA

P. GADSON, Department of Anatomy, Medical College of Georgia, Augusta,
    GA, USA

W. GALLWITZ, University of Texas Health Science Center, Department of
    Obstetrics & Gynecology, San Antonio, TX, USA

R.M. GARDNER, Department of Pharmacology, The University of Texas Medical
    School, Houston, TX 77225, USA

D.J. GRUOL, Department of Regulatory Biology, The Salk Institute, 10010
    North Torrey Pines Road, LaJolla, CA 92037, USA

M.T. HARRIGAN, Department of Regulatory Biology, The Salk Institute, 10010
    North Torrey Pines Road, LaJolla, CA 92037, USA

R.A. HIIPAKKA, The Ben May Institute, 950 East 59th Street, Box 424,
    University of Chicago, Chicago, IL 60637, USA

Y.-P. HWUNG, Department of Cell Biology, Baylor College of Medicine, 1
    Baylor Plaza, Houston, TX 77030, USA

C. JEWELL, Department of Physiology, University of North Carolina, Chapel
    Hill, NC 27599, USA

D.B. JUMP, Department of Physiology, Michigan State University, 214 Giltner
    Hall, East Lansing, MI 48824, USA

M. KARIN, Department of Pharmacology M-036, University of California-San
    Diego, School of Medicine, LaJolla, CA 92093, USA

D. KETTELBERGER, J.M. KIM, University of Texas Health Science Center,
    Department of Cellular and Structural Biology, San Antonio, TX, USA

J.L. KIRKLAND, Division of Endocrinology, Department of Pediatrics, Baylor
    College of Medicine, Houston, TX, 77030, USA

G. KLOCK, Institute of Cell and Tumor Biology, German Cancer Research
    Center, 6900 Heidelberg 1, Federal Republic of Germany

H. LERIVRAY, National Institute for Medical Research, The Ridgeway, Mill
    Hill, London NW7 1AA, England

S. LIAO, The Ben May Institute, 950 East 59th Street, Box 424, University of
    Chicago, Chicago, IL 60637, USA

T.-H. LIN, Division of Endocrinology, Department of Pediatrics, Baylor College
    of Medicine, Houston, TX 77030, USA

R.B. LINGHAM, Department of Pharmacology, The University of Texas Medical
    School, Houston, TX 77225, USA

D.S. LOOSE-MITCHELL, Department of Pharmacology, The University of Texas
    Medical School, Houston, TX 77225, USA

M.A. MANCINI, University of Texas Health Science Center, Department of
    Cellular and Structural Biology, San Antonio, TX, USA

J. MARSH, National Institute for Medical Research, The Ridgeway, Mill Hill,
    London NW7 1AA, England

S.C. MARTIN, National Institute for Medical Research, The Ridgeway, Mill
    Hill, London NW7 1AA, England

B.M. MARKAVERICH, Department of Cell Biology, Baylor College of Medicine,
    Houston, TX 77030, USA

R. MESTRIL, Institute of Cell and Tumor Biology, German Cancer Research
    Center, 6900 Heidelberg 1, Federal Republic of Germany

B.S. MIDDLEDITCH, Department of Biochemical and Biophysical Sciences,
    University of Houston, Houston, TX 77004, USA

V.R. MUKKU, Department of Pharmacology, The University of Texas Medical
    School, Houston, TX 77225, USA

B.W. O'MALLEY, Department of Cell Biology, Baylor College of Medicine,
    Texas Medical Center, Houston, TX 77030, USA

D.H. OH, University of Texas Health Science Center, Department of Obstetrics & Gynecology, San Antonio, TX, USA

C. ORENGO, Department of Pharmacology, The University of Texas Medical School, Houston, TX 77225, USA

A.K. ROY, Division of Molecular Genetics, Department of Obstetrics & Gynecology, University of Texas Health Science Center, 7703 Floyd Curl Drive, San Antonio, TX 78284, USA

W. SCHMID, Institute of Cell and Tumor Biology, German Cancer Research Center, 6900 Heidelberg 1, Federal Republic of Germany

G. SCHULTZ, Institute of Cell and Tumor Biology, German Cancer Research Center, 6900 Heidelberg 1, Federal Republic of Germany

C. SILVA, Department of Biochemistry, University of North Carolina, Chapel Hill, NC 27599, USA

S.S. SIMONS, JR., Steroid Hormones Section, NIDDK/LAC, Building 8, Room B2A-07, National Institutes of Health, Bethesda, MD 20892, USA

J. SIMENTAL, Department of Physiology, University of North Carolina, Chapel Hill, NC 27599, USA

C.S. SONG, University of Texas Health Science Center, Department of Cellular and Structural Biology, San Antonio, TX, USA

G.M. STANCEL, Department of Pharmacology, University of Texas Medical School, P.O. Box 20708, Houston, TX 77025, USA

U. STRAHLE, Institute of Cell and Tumor Biology, German Cancer Research Center, 6900 Heidelberg 1, Federal Republic of Germany

J.R. TATA, National Institute for Medical Research, The Ridgeway, Mill Hill, London NW7 1AA, England

L.E. THEILL, Department of Pharmacology M-036, University of California-San Diego, School of Medicine, LaJolla, CA 92093, USA

E.B. THOMPSON, Department of Human Biological Chemistry and Genetics, 226 Basic Science Bldg. F36, The University of Texas Medical Branch, Galveston, TX 77550, USA

S.Y. TSAI, Department of Cell Biology, Baylor College of Medicine, 1 Baylor Plaza, Houston, TX 77030, USA

M-J. TSAI, Department of Cell Biology, Baylor College of Medicine, Texas Medical Center, Houston, TX 77030, USA

D. TULLY, Department of Biochemistry, University of North Carolina, Chapel Hill, NC 27599, USA

Y.M. WANG, Curriculum in Neurobiology, University of North Carolina, Chapel Hill, NC 27599, USA

L.-H. WANG, Department of Cell Biology, Baylor College of Medicine, 1 Baylor Plaza, Houston, TX 77030, USA

G. WASNER, Institut fur Molekularbiologie, Billrothstrasse 11, A-5020 Salzburg, Austria

# TOWARDS A MOLECULAR UNDERSTANDING OF STEROID BINDING
## TO GLUCOCORTICOID RECEPTORS

S. Stoney Simons, Jr.

Introduction

Steroid hormone binding to receptor proteins in target cells is the first step in a series of events that translate the structural information of the steroid into the observed biological responses. The precise type of activity (e.g., agonist, antagonist, or a mixture of the two) elicited by a given steroid can vary with the gene examined, even in the same cell (Mercier et al., 1986; Simons et al., 1988a and 1988b). Nevertheless, all of the information that is transmitted by the receptor-steroid complex for the expression of agonist vs. antagonist activity is necessarily contained in the structure of the steroid, which somehow gives rise to receptor-steroid complexes with different biochemical and biological properties.

Despite this pivotal role of steroid-receptor interactions, the molecular details are almost completely unknown. A rather large (≈250 amino acids) steroid binding domain for all steroid receptors has been defined (Giguere et al., 1986; Kumar et al., 1986; Godowski et al., 1987; Arriza et al., 1987). There had been only indirect chemical evidence that the steroid binding site contained a cysteine (Baxter and Tomkins, 1971; Simons and Thompson, 1982; Harrison et al., 1983; Bodwell et al., 1984b; Formstecher et al., 1984) until just recently (Simons et al., 1987; Carlstedt-Duke et al., 1988; Smith et al.,1988). Similar indirect evidence suggested that lysine (DiSorbo et al., 1980; O'Brien and Cidlowski, 1980; Naray and Bathori, 1982) and arginine (DiSorbo et al., 1980) may additionally be involved in steroid binding to glucocorticoid receptors.

In order to obtain further information about the molecular interactions of steroids with glucocorticoid receptors, we wanted to specifically modify the receptor protein in a manner that caused minimal alterations in the tertiary structure and preserved as much biological function as possible. Thus we could not use molecular biology approaches, such as point mutagenesis and deletion mutants, since there is currently no way of predicting the effect of such modifications on the tertiary structure of proteins. Instead, we decided to use chemical reagents to selectively modify specific amino acids of the receptor protein. With this approach, we should be able to determine some of the amino acids that are involved in steroid binding and eventually which molecular interactions are responsible for what type of biological activity. In this chapter, I will discuss how we have used two thiol-specific reagents to identify the involvement of two different cysteines in the binding of steroids to glucocorticoid receptors. The first reagent is the glucocorticoid receptor affinity label dexamethasone 21-mesylate (Dex-Mes) (Simons and Thompson, 1981; Eisen et al., 1981). The second reagent is the sterically small compound, methyl methanethiolsulfonate (MMTS) (Smith et al., 1975; Brocklehurst, 1979).

## Experimental and Discussion

### Dex-Mes

Dex-Mes has been established as an irreversible antiglucocorticoid (Simons and Thompson, 1981; Simons and Miller, 1986; Simons et al., 1988a) which specifically labels glucocorticoid receptors in broken (Simons and Thompson, 1981; Eisen et al., 1981; Simons, 1988) and whole cell (Simons et al., 1983; Simons, 1988) preparations. This labeling appears to occur in the steroid binding cavity since (1) Dex-Mes labels 98 ± 11% of the available receptors (Simons and Miller, 1984) and this labeling is prevented by excess dexamethasone (Dex) (and *visa versa*; Simons and Thompson, 1981; Eisen et al., 1981), (2) Dex-Mes inhibits all of the binding of [$^3$H]Dex to receptors (Simons and Thompson, 1981; Eisen et al., 1981), (3) covalently labeled receptors display normal activation (Simons et al., 1983; Simons and Miller, 1984) and DNA binding (Simons and Miller, 1984; Miller et al., 1984; Miller and Simons, 1988a), and (4) receptors labeled in intact cells display nuclear translocation (Simons et al., 1983; Simons and Miller, 1986; Sistare et al., 1987; Miller and Simons, 1988a).

Specificity of Reactions of Dex-Mes: The early conclusions that the steroid binding cavity of glucocorticoid receptors contained cysteine, lysine, and arginine (see above) were based on the observations that various reagents, thought to be specific for sulfhydryl or amino groups, blocked the subsequent binding of glucocorticoids to the receptor. However, it is difficult to rule out indirect effects on non-receptor components or indirect steric effects on the tertiary structure of the receptor by bulky reagents in these experiments. Furthermore, some of the less bulky reagents (e.g., N-ethylmaleimide and iodoacetamide) are not completely specific (see references in Bodwell et al., 1984a), or at least their specificity has not been established under the reaction conditions used with receptors. For these reasons, we first needed to rigorously establish the reactivity of Dex-Mes before attempting to identify the amino acids of the glucocorticoid receptor that are covalently labeled by Dex-Mes.

Our initial studies indicated that Dex-Mes selectively and rapidly reacted with ionized thiol groups in acetone solutions at 0°C (Simons et al.,1980). In addition, the covalent labeling of total cytosolic protein increased at pHs above 7.0 and was prevented by the presence of 50 mM $\beta$-mercapto-ethanol (Simons et al.,1983). Further studies were conducted with bovine serum albumin (BSA+SH), which has only one free -SH group at Cys-34 (Brown, 1977), as a model protein. For comparison, we used a chemically modified BSA (BSA-SH) which has no -SH groups due to the complete conversion of Cys-34 to a mixed disulfide. Initial experiments indicated that the labeling of BSA+SH with [³H]Dex-Mes increased with time ($T_{1/2} \approx 4$ hr at 22°C), temperature, and pH (a steady increase up to pH 8.9 at 22°C) (Simons, 1987). At early times of reaction with Dex-Mes (i.e., 2 hr), about 90% of the labeling of BSA+SH was at Cys-34. When $4.6 \times 10^{-6}$ M BSA ± SH was preincubated with $9.6 \times 10^{-6}$ M of the thiol specific reagent MMTS for 90 min at 24°C, the labeling of BSA+SH was reduced to the level seen for BSA-SH while the amount of labeled BSA-SH was not altered (Simons, 1987). Furthermore, protease digestion of [³H]Dex-Mes labeled BSA+SH gave those ³H-labeled bands on SDS-polyacrylamide gels that would be expected if only Cys-34 had been labeled (Simons, 1987). These data argue that Dex-Mes (and also MMTS) selectively reacts with the cysteine -SH group of proteins in basic aqueous solutions.

Dex-Mes Labeling of Sulfhydryl Groups in the Glucocorticoid Receptor:
Dex-Mes labeling of the intact 98K daltons receptor, and the 82K daltons receptor fragment (Reichman et al., 1984), in crude HTC cell cytosol solutions is virtually eliminated by a 2.5 hr preincubation with $10^{-4}$ M MMTS at 0°C. Furthermore, increasing concentrations of MMTS reduce Dex-Mes labeling of, and Dex binding to, receptors equally (Simons, 1987). We therefore conclude that Dex-Mes labeling of HTC cell glucocorticoid receptors occurs at the same free -SH group(s) that is involved in Dex binding to the receptor.

In order to determine how many of the 20 -SH groups present in the rat glucocorticoid receptor (Miesfeld et al., 1986) are covalently modified,

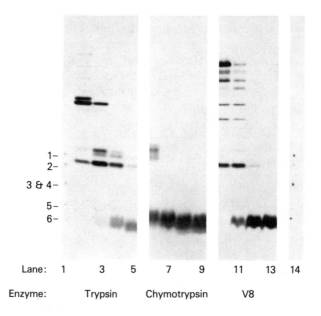

**Figure 1:** Fluorograph of limit protease digestions of activated, [³H]Dex-Mes labeled, HTC cell glucocorticoid receptors. Activated complexes were denatured (0.2% SDS/0.5 mM DTT for 30 min at 22°C) and digested for 4 min and 1,4 and 24 hr (for lanes 2-5 and lanes 6-9), or for 1,3,7 and 24 hr (for lanes 10-13), and analyzed on 15% SDS-polyacrylamide gels, which were dried and fluorographed. The sizes of the molecular weight markers (numbered 1-6 = cyanogen bromide fragments of myoglobin) are 17,200, 14,600, 8240, 6380, 2560, and 1695 daltons. (from Simons, 1987)

Dex-Mes labeled receptors were first partially purified by DNA-cellulose chromatography (Reichman et al., 1983). This procedure gave preparations in which the only [³H]Dex-Mes labeled protein was the receptor. Progressive protease digestion of these partially purified, labeled receptors gave a series of Dex-Mes labeled fragments which could be easily separated on SDS-polyacrylamide gels (Figure 1).

Quantitation of the radioactivity present in the smallest $M_r$ fragment generated by 24 hr of digestion with each protease indicated that all of the [³H]Dex-Mes initially present in the 98,000 daltons receptor was recovered in the limit fragments of V8 protease and chymotrypsin digestion (Simons, 1987). This demonstrated that there are no other, smaller [³H]Dex-Mes labeled fragments which did not remain on the gel and which would have derived from a second site on the receptor that was labeled by [³H]Dex-

Mes. We interpreted the approximately 50% recovery of the radioactivity in the trypsin limit digest fragment as indicating the presence of a second, internal trypsin site, at which only partial cleavage occurs, to give peptides that are too small to remain on the gel during staining and destaining (Simons, 1987; see also below). These results argue that each [3]H-labeled limit digest band corresponds to a single peptide containing one or more labeled cysteines.

Identification of Dex-Mes Labeled Cysteines in the Glucocorticoid Receptor: While the above limit protease digestion solutions contained many peptides, only those species derived from the intact, affinity labeled 98K daltons receptor contained [[3]H]Dex-Mes (Figure 1). Thus radiochemical monitoring of an Edman degradation of these limit digest solutions would provide information about those portions of the receptor that contain covalently attached [[3]H]Dex-Mes. To prepare the limit digests for this analysis, each solution was first chromatographed on a Sep-pak $C_{18}$ cartridge in order to change the solvent to 90% MeOH/10% water/ 0.1% trifluoroacetic acid. In some cases, further chromatography on a $C_3$ HPLC column was performed but this was not necessary.

Analysis of the first 20 cycles in the sequential Edman degradation of the $C_{18}$ Sep-pak eluted 1.8K daltons V8 protease limit digest fragment, either before (data not shown) or after (Figure 2A) subsequent HPLC chromatography, revealed that most of the released radioactivity was in cycle 7. The tailing of the radioactive peak into the following fractions is normal and is due to out-of-phase degradation. The sum of radioactivity observed in cycles 7-9 indicates that >86% of the covalently bound [[3]H]steroid is attached to a single cysteine. With the trypsin limit digest material, essentially all of the radioactivity in the first 20 cycles was found in cycle 5 (Figure 2B). There are three cysteines in the HTC cell glucocorticoid receptor (Miesfeld et al., 1986) which are five amino acids downstream from a basic amino acid residue that could be cleaved by trypsin. Of these three cysteines, only Cys-656 is also compatible with the V8 protease limit digest data of Figure 2A (see also Figure 3).

Sequential Edman degradation was also performed on the chymotrypsin limit digest material to give one major peak of radioactivity in cycle 17 (Figure 2C). There are four cysteines that are 17 amino acids downstream from a conventional chymotrypsin cutting site (i.e., the aromatic amino acids phenylalanine, tyrosine, or tryptophan and leucine [Kasper, 1985]). In all cases, the cleavage sites for chymotrypsin would be a leucine. Of these four cysteines, only Cys-656 is also compatible with the trypsin data. Collectively, the sequential Edman degradation data for the trypsin, chymotrypsin, and V8 protease limit digests uniquely implicate Cys-656 as

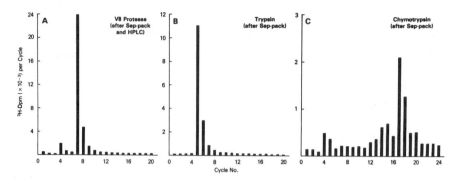

**Figure 2:** Sequential Edman degradation of limit protease digestion fragments of activated, [³H]Dex-Mes labeled, HTC cell glucocorticoid receptors. Activated, labeled receptors were denatured as in Figure 1, digested with V8 protease, trypsin, or chymotrypsin at 150 μg/ml for 16-18 hr at 22°C, and chromatographed on $C_{18}$ Sep-pak cartridges. The peak fractions from the Sep-pak cartridges were subjected to sequential Edman degradation on an Applied Biosystems Model 470A protein/peptide sequencer as previously described (Rudikoff et al., 1986). The dpm per Edman degradation cycle were plotted for **A)** a $C_{18}$ Sep-pak chromatographed, and then $C_3$ HPLC fractionated, V8 protease limit digest, **B)** a $C_{18}$ Sep-pak chromatographed, trypsin limit digest, and **C)** a $C_{18}$ Sep-pak chromatographed, chymotrypsin limit digest. In each case, the complete conversion of the 98K daltons receptor to the limit protease digestion fragment was demonstrated by electrophoresis on 15% SDS-polyacrylamide gels and detection by fluorography (data not shown). (from Simons et al., 1987)

the single cysteine of the HTC cell glucocorticoid receptor that is affinity labeled by Dex-Mes. The location of the observed and predicted protease digestion sites around Cys-656 are shown in Figure 3.

While this conclusion relies on the above evidence that Dex-Mes labeling of the receptor occurs exclusively at a cysteine, the same conclusion can be obtained without initially knowing the chemical identity of the labeled residue. A computer search of the rat glucocorticoid sequence (Miesfeld et al., 1986) for an amino acid that is 17 residues downstream from a chymotrypsin site, 7 from a V8 protease site, and 5 from a trypsin site yields Asp-122 and Cys-656 as the only possibilities (Simons et al., 1987). However, Asp-122 can be eliminated because mild chymotrypsin digestion of Dex-Mes labeled receptors yields a 42K daltons fragment which retains the covalently bound steroid (Reichman et al., 1983) but is lacking Asp-122

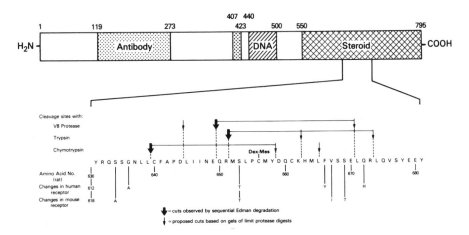

<u>Figure 3:</u> Protease cleavage sites near [$^3$H]Dex-Mes labeled Cys-656 of the HTC cell glucocorticoid receptor. The various domains of the receptor (antibody binding sites, DNA binding domain, and steroid binding domain) in the intact receptor (Miesfeld et al., 1986) are shown on top. Antibodies 250 (Okret et al., 1984) and BUGR2 (Gametchu and Harrison, 1984) recognize amino acids contained between 119-273 and 407-423 respectively (Rusconi and Yamamoto, 1987). Below is given the single letter amino acid sequence, and numbering, of the region surrounding Cys-656 of the HTC cell receptor (Miesfeld et al., 1986). The heavy arrows (⬇) indicate the cleavage sites of V8 protease, trypsin, and chymotrypsin that were identified by sequential Edman degradation. The light arrows (↓) show the probable additional cleavage sites that would account for the observed sizes of the single protease limit digest fragments (indicated by ———) and the other small species seen after single and double limit digests (see Simons et al., 1987). Those amino acids which are different in the human (Hollenberg et al., 1985) and mouse (Danielson et al., 1986) glucocorticoid receptor are indicated. (from Simons et al., 1987)

(Miesfeld et al., 1986; Simons et al., 1987). This identification of Cys-656 as the amino acid labeled by Dex-Mes has recently been confirmed by others for rat (Carlstedt-Duke et al., 1988) and mouse (Smith et al., 1988) glucocorticoid receptors (c.f., Figure 2).

MMTS

<u>Specificity of Reactions of MMTS:</u> Previous studies by other investigators

(Smith et al., 1975; Brocklehurst, 1979), in addition to the data presented above, show that MMTS reacts almost exclusively with the -SH groups of proteins in basic aqueous solutions. An added benefit of MMTS is that its small size permits reaction with relatively inaccessible sulfhydryl groups. Furthermore, the indirect steric effects on a protein that has been treated with MMTS are minimal since the van der Waals interaction volume of the added methylthiol (-S-CH$_3$) group is smaller than that of the added groups of almost all other thiol reagents.

Effect of MMTS on [$^3$H]Dex Binding and [$^3$H]Dex-Mes Labeling of Receptors: The ability of thiol reagents, such as iodoacetamide, N-ethylmaleimide , and N-phenylmaleimide, to inhibit steroid binding to glucocorticoid receptors is well known (Baxter and Tomkins, 1971; Koblinsky et al., 1972; Rees and Bell, 1975). Recently, we have shown that these thiol reagents totally inhibit the subsequent binding of [$^3$H]Dex to cell-free receptors at reagent concentrations that are only slightly higher than the ≈0.8 mM thiol concentration of the receptor solutions. The reactions with receptor are rapid (≤30 min at 0°C), give nice sigmoidal dose-response curves, and occur even in the presence of 20 mM molybdate (Miller and Simons, 1988b), which is known to interact with thiols (Kaul et al., 1985; Garner and Bristow, 1985).

The shape of the dose-response curve for MMTS inhibition of [$^3$H]Dex binding to crude HTC cell receptors (Figure 4) is very different from that of the above thiol reagents. With no preincubation with MMTS, a biphasic curve is obtained. After a 2.5 hr preincubation with MMTS, the curve becomes bimodal in that low and high, but not intermediate, concentrations of MMTS eliminate [$^3$H]Dex binding (Figure 4). The effect of time, however, is appreciable only with MMTS concentrations near $10^{-4}$M (Figure 4A). The disappearance of steroid binding at 100 mM MMTS appears to be due to the denaturation of protein; a reaction with thiol groups is unlikely since this concentration of MMTS is 100 times that of the thiol content of the cytosol solution. Thus two thiols, or groups of thiols, (one of which reacts at ≈0.1 mM MMTS and another that reacts at 1-10 mM MMTS) appear to influence the initial binding of steroid to receptors. It should be noted that steroid bound receptors (either unactivated or activated) undergo little or no dissociation after treatment with 0.3 to 30 mM MMTS for 2 $_{1/2}$ hr at 0°C (Miller and Simons, 1988b).

The bimodal curve for [$^3$H]Dex binding that is seen after a 2.5 hr MMTS preincubation (Figure 4B) is not obtained for [$^3$H]Dex-Mes labeling of Cys-656 (Figure 5A). MMTS reaction with -SH groups is very rapid in that [$^3$H]Dex-Mes labeling of HTC cell receptors and cytosolic proteins are completely inhibited by coincubation with 3 mM MMTS (Miller and Simons,

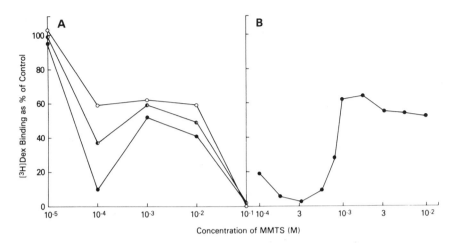

Figure 4: Inhibition of [³H]Dex binding to HTC cell receptors by preincubation with MMTS. **A**) Effect of MMTS preincubation time on subsequent [³H]Dex binding. Duplicate samples of HTC cell cytosol solution were pretreated with various concentrations of MMTS in absolute EtOH for 0 hr ( O ), 0.5 hr ( ◑ ), or 2.5 hr ( ● ) before the addition of [³H]Dex ± excess [¹H]Dex. After a further 2.5 hr incubation, the specific binding to receptors was determined after adding dextran-coated charcoal to remove free steroid and plotted against the concentration of added MMTS. The data points shown are the average values derived from 1-5 experiments. **B**) Detailed does-response curve for inhibition of [³H]Dex binding after 2.5 hr preincubation with MMTS. Duplicate samples of HTC cell cytosol solution were treated with various concentrations of MMTS for 2.5 hr before determining the remaining steroid binding activity of the receptors as above in A. (from Miller and Simons, 1988b)

1988b). Nevertheless, the length of receptor preincubation with 0.3 mM MMTS does influence the subsequent amount of [³H]Dex-Mes labeling of receptors (Figure 5B). Collectively, the Dex binding and Dex-Mes labeling data indicate a very fast reaction of -SH groups at all MMTS concentrations but an ensuing, slower reaction for ≈$10^{-4}$M MMTS pretreated receptors which initially retain some free -SH groups (see Figure 5B).

Further studies revealed that a 10-fold dilution of reactants produced less than a 2-fold decrease in the kinetics of loss of steroid binding following $10^{-4}$ M MMTS treatment of steroid-free receptors (Miller and Simons, 1988b). Thus this slow loss of steroid binding involves an intramolecular reaction that is unaffected by dilution.

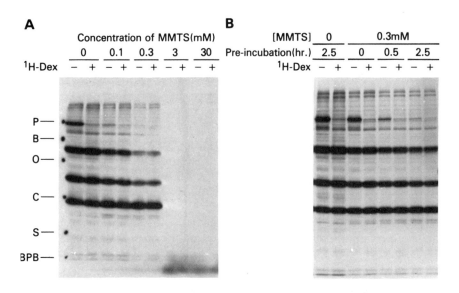

**Figure 5:** Inhibition of [³H]Dex-Mes labeling of HTC cell receptors and cytosolic proteins by preincubation with MMTS. **A)** Dose-response for inhibition of [³H]Dex-Mes labeling after 2.5 hr preincubation with MMTS. Aliquots of HTC cell cytosol solution were treated with the indicated MMTS concentrations. After a 2.5 hr incubation, aliquots were removed to determine [³H]Dex binding as in Figure 4 and to be labeled by [³H]Dex-Mes ± [¹H]Dex for 2.5 hr. The reactions with [³H]Dex-Mes were stopped by quick freezing at -78°C and then electrophoresed on 10.8% SDS-polyacrylamide gels, which were fluorographed to give the above film. **B)** Effect of MMTS preincubation time on inhibition of labeling by [³H]Dex-Mes. Samples were prepared, preincubated with 0.3 mM MMTS, and then assayed for [³H]Dex-Mes labeling as above in A.

For the experiments of A and B, [³H]Dex binding was virtually identical to that of Figure 4. P,B,O,C, and S identify the molecular weight markers (M_r = 97,400, 66,300, 45,000, 30,600, 21,500, and 14,400 respectively; BPB = bromphenol blue); the lines point to the spots of UltEmit that were overlaid on the molecular weight markers. The radioactivity running ahead of the dye front in samples treated with ≥3 mM MMTS is probably due to an abundant small molecule(s) labeled by Dex-Mes at functional groups other -SH (Simons, 1987). (from Miller and Simons, 1988b)

Properties of MMTS-modified, steroid-free receptors. Table 1A shows
that $10^{-5}$M MMTS pretreated receptors can be converted to either the non-
binding or the binding form with increasing concentrations of MMTS. In
contrast, added MMTS has no effect on the non-binding, $10^{-4}$ M MMTS
pretreated receptors. Thus the non-binding and binding forms of receptor
that are obtained at 0.1-0.3 and $\geq 1$ mM MMTS respectively are not
produced sequentially. This, in turn, argues that the same thiols react at
low and intermediate MMTS concentrations but that the reaction products
are two different chemical species.

The loss of [$^3$H]Dex-Mes labeling of any of the usually labeled proteins
after incubation with $\geq 3$ mM MMTS (Figure 5A) argues that these solutions
no longer contain free -SH groups. Further evidence for this comes from
the observation that those concentrations of iodoacetamide which
completely block steroid binding to receptors have little or no ability to
reduce the binding of steroid to 2 mM MMTS pretreated receptors (Table
1B). A further conclusion from these data is that Dex binding to
glucocorticoid receptors can occur in the absence of any free -SH groups.

All of the above experiments were conducted in the presence of 20 mM
$Na_2MoO_4$. Scatchard analyses show that the affinity of receptors
pretreated with 3 mM MMTS $\pm$ 20 mM $Na_2MoO_4$ for [$^3$H]Dex is reduced by
80% (Miller and Simons, 1988b). However, the remaining binding capacity
is much higher in the presence of molybdate (i.e., $\approx$70% vs. $\approx$15%). This
difference occurs because 20 mM $Na_2MoO_4$ prevents much of the loss of
steroid binding that is seen in the first few minutes of exposure to MMTS
(Miller and Simons, 1988b).

Reversibility of MMTS Reactions with the Glucocorticoid Receptor: Since
the reactions of MMTS are specific for thiols, the effects of MMTS
preincubation with glucocorticoid receptors should be readily reversed by
the addition of thiols such as dithiothreitol (DTT). In fact, the inhibition of
[$^3$H]Dex binding by 0.3 mM MMTS preincubation for 2.5 hr in the presence
of $Na_2MoO_4$ is completely reversed by increasing concentrations of DTT
(Figure 6A). The reversal of the effects of preincubation with 3 mM MMTS
plus $Na_2MoO_4$ is more complex (Figure 6B). At low concentrations of DTT,
there is a decrease in steroid binding. This decrease is more pronounced
with longer periods of DTT incubation. With higher DTT concentrations,
there is a rapid and nearly complete recovery of steroid binding capacity.
However, the inhibition of steroid binding to receptors pretreated with 3 mM
MMTS in the absence of molybdate could not be effectively reversed either
by up to 2.5 hr of DTT or by 0.5 hr of DTT + $Na_2MoO_4$; a total of 2.5 hr of
DTT + $Na_2MoO_4$ was required for complete reversal (Miller and Simons,
1988b). In contrast, molybdate had no effect on the reversibility of 0.3 mM

## Table 1

### Effect of Sequential Thiol Reagent Preincubation on [3H]Dex Binding to Steroid-free Receptors

**A)**

| First Pre-incubation | Second Pre-incubation | Percent of Control Binding After Second Preincubation at Concentration (M) of | | | | | |
|---|---|---|---|---|---|---|---|
| | | 0 | $10^{-5}$ | $10^{-4}$ | $10^{-3}$ | $10^{-2}$ | $10^{-1}$ |
| - | MMTS | 100 | 89 | 8 | 58 | 42 | 1 |
| $10^{-5}$M MMTS | MMTS | | 98 | 7 | 48 | 45 | 3 |
| $10^{-4}$M MMTS | MMTS | | | 2 | 3 | 2 | 0 |

**B)**

| First Pre-incubation | Second Pre-incubation | 0 | $3\times10^{-4}$ | $10^{-3}$ | $2\times10^{-3}$ | $10^{-2}$ | $2\times10^{-2}$ | $10^{-1}$ |
|---|---|---|---|---|---|---|---|---|
| - | IA | 100 | (40) | | 1 | | 0 | |
| - | MMTS | 100 | | 73 | (56) | 60 | | 0 |
| $3\times10^{-4}$M IA | MMTS | 40 | | 20 | | 15 | | 0 |
| $2\times10^{-3}$M MMTS | IA | 56 | | | 51 | | 39 | |

**A)** Sequential preincubations with MMTS were conducted by pretreating HTC cell cytosol solutions with 0, $10^{-5}$, or $10^{-4}$M MMTS for 2.5 hr. Quadruplicate 95 µl aliquots were then treated with 1 µl of EtOH ± MMTS to give the indicated total MMTS concentrations of the second preincubation. After an additional 2.5 hr incubation, duplicate samples were assayed for steroid binding as described in Figure 4 and expressed as percentage of the EtOH/EtOH preincubated control. All data represent the average of two experiments except for the one experiment where the first preincubation was with $10^{-5}$M MMTS. **B)** Sequential preincubation with iodoacetamide (IA) and MMTS were conducted as in (A) except that the first preincubation of HTC cell cytosol (15% by volume) was for 0.5 hr. After the second preincubation (2.0 to 2.5 hr), and binding with [3H]Dex ± [1H]Dex for 2.5 hr, the average specific binding was determined and expressed percent of the EtOH/EtOH control. The numbers in parentheses are values obtained when the indicated concentration of IA or MMTS was present during the first and second preincubation as opposed to only during the second preincubation. (from Miller and Simons, 1988b).

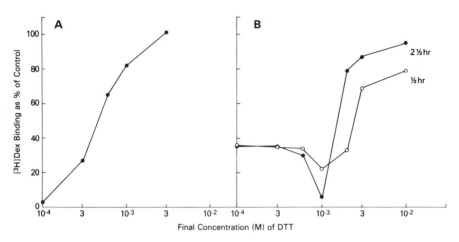

Figure 6: DTT reversal of MMTS inhibition of [$^3$H]Dex binding to steroid-free receptors. **A**) Reversal of 3 x 10$^{-4}$M MMTS preincubation. Duplicate tubes of HTC cell cytosol solution (94 µl) were preincubated with 1% EtOH ± 3 x 10$^{-6}$M for 2.5 hr. The EtOH control was then treated with 1 µl of pH8.8 TAPS buffer while other samples received 1 µl of 100x DTT in pH8.8 TAPS buffer. After 0.5 hr of incubation, 4 µl of [$^3$H]Dex ± [$^1$H]Dex was added for 2.5 hr. The average specific binding to receptors was determined as described in Figure 4, expressed as percentage of EtOH control binding, and plotted versus the final DTT concentration. **B**) Reversal of 1.6 x 10$^{-3}$M MMTS preincubation. Control tubes of HTC cell cytosol solution were preincubated in duplicate for 2.5 hr with 1 µl of EtOH, followed by another 2.5 hr with 1 µl of pH8.8 TAPS buffer. The remaining samples were preincubated in batch with 1.6 x 10$^{-3}$M MMTS in EtOH for 2.5 hr, followed by 1 µl of 100x DTT in pH8.8 TAPS buffer for 0.5 or 2.5 hr. The receptor binding activity of all samples, plus control, was then assayed by adding [$^3$H]Dex ± [$^1$H]Dex. After a 2.5 hr incubation, the average specifically bound [$^3$H]Dex was determined and plotted as above in A for samples incubated with DTT for 0.5hr (O) and for 2.5 hr (●). (from Miller and Simons, 1988b)

MMTS pretreated receptors (Miller and Simons, 1988b). These data suggest that both molybdate, and the proposed intramolecular disulfide formed at 0.3 mM MMTS, is able to maintain oxidized receptors in a conformation that rapidly recovers steroid binding activity upon the addition of DTT.

Model of MMTS Modifications of the Glucocorticoid Receptor: On the

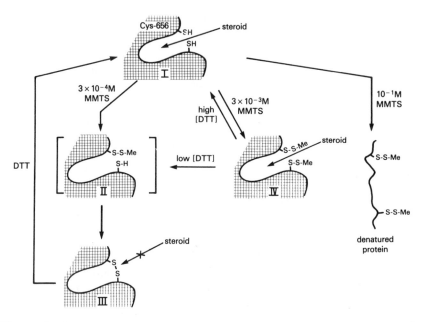

**Figure 7:** Model of MMTS/DTT modifications of sulfhydryl groups in the steroid binding cavity of glucocorticoid receptors. A hypothetical steroid binding cavity of the receptor containing part of the receptor protein and Cys-656 is shown. Steroid can bind to the fully reduced, unmodified receptor (I) and to the mixed disulfide form (IV) but not to the intramolecular disulfide form (III). (from Miller and Simons, 1988b)

basis of the above data, we propose a model for steroid binding to rat glucocorticoid receptors that involves two -SH groups, one of which would be Cys-656 (Figure 7). This model explains all of our observed results. (1) The transition between no binding and steroid binding after MMTS preincubation occurs when there is enough MMTS ($\approx 1$ mM in Figure 4) to react with all of the accessible -SH groups ($\approx 0.8$ mM by Ellman's reagent). (2) The size of the functional group that is transferred to the receptor (I) by MMTS is smaller than that by other thiol reagents. Consequently steroids can still bind (albeit with reduced affinity) to receptors in which both thiols have been modified by MMTS (i.e., IV) but not to receptors modified by bulkier thiol reagents, such as iodoacetamide. (3) The unique ability of proteins containing some MMTS-modified thiol groups to undergo disulfide exchange (e.g., II to give III) explains both the loss of steroid binding at MMTS concentrations below the cytosol -SH concentration (see Figures 4 and 5) and the slower kinetics for the effect of low *vs.* intermediate MMTS concentrations (see Figure 4A). In this context, MMTS acts as a zero-length cross-linker. (4) Only an intramolecular reaction for this slow step at low MMTS concentrations accounts for the insensitivity of rate to changes in cytosol concentration. (5) Receptors reacted with 0.3 and 3 mM MMTS represent two, distinct species (III and IV in Figure 7) in which both thiols have been differently modified. Therefore treatment of 0.3 mM MMTS modified receptors (III) with 3 mM MMTS is without effect (Table 1). (6) The model predicts exactly the kinetics and bimodal behavior seen for DTT reversal of the effects of preincubation with ≈2 mM MMTS (Figure 6B). Thus receptors pretreated with ≈2 mM MMTS (IV) can be converted to the non-binding form (II) with low concentrations of DTT while the converse is not possible (see item #5). (7) Dex-Mes labeling of the receptor is lost (Figure 5), while both Dex (Figure 4) and Dex-Mes (Miller and Simons, 1988b) binding are retained, because ≥1 mM MMTS reaction with both Cys-656 and the second thiol to give IV in Figure 7 will prevent affinity labeling but not steroid binding. (8) Large amounts of iodoacetamide do not decrease the steroid binding of 3 mM MMTS pretreated receptors (Table 1B) because all of the accessible thiols have already been modified by MMTS. Therefore, in contrast to the current dogma (Mendel et al., 1987; Sanchez et al., 1987b), we conclude that steroids can bind to glucocorticoid receptors in which all of the accessible thiols are no longer reduced *if the modifying group is sterically small* (e.g., $-S-CH_3$ *vs.* $-S-CH_2-CONH_2$).

We strongly suspect that Cys-656 is one of the thiols in the model of Figure 7 since it is known to be in the steroid binding cavity (Simons et al., 1987) , is probably near the opening of the steroid binding cavity (Simons et al., 1987; Pons et al., 1985), and appears to react with MMTS to become inaccessible for Dex-Mes labeling under conditions where most of the other cytosolic -SH groups do not react (Figure 5). We have depicted the other thiol as being one of the other 19 thiols of the rat receptor (Miesfeld et al., 1986), such as Cys-754 [which is photoaffinity labeled by triamcinolone acetonide (Carlstedt-Duke et al., 1988)], mostly for convenience. The above kinetic experiment requires that the disulfide III of Figure 7 be formed by an intramolecular reaction. However, these kinetic results are also totally consistent with the second thiol residing on a non-receptor molecule that is non-covalently associated with the receptor, such as hsp90 (Sanchez et al., 1985, 1987a), a 72K daltons protein (Wrange et al., 1984), or a 59K daltons protein (Tai et al., 1986).

We do not yet know how the second thiol normally participates in steroid binding to the glucocorticoid receptor. Part of this answer, plus the generality of this second thiol in the binding of steroid to other receptors, should be evident if our current attempts to locate this second thiol are successful. In view of the apparent proximity of the second thiol to Cys-656, which reacts with the C-21 position of Dex-Mes, this other thiol may hydrogen bond to the polar substituents of the C-17 position of most glucocorticoids and antiglucocorticoids. Alternatively, the ability of this second thiol to reversibly form an intramolecular disulfide may be invoked in times of metabolic stress, of modulation of receptor levels, or of receptor synthesis, to give a form of the receptor that will not bind steroid but can rapidly yield functional receptors upon reduction of the disulfide linkage. In this respect, it is of interest that the presence of the oxidative phosphorylation inhibitor dinitrophenol causes the whole cell formation of a class of nuclear glucocorticoid receptors that do not bind steroid (Mendel et al., 1986).

## Conclusions

Our studies with two chemically specific, reactive molecules (Dex-Mes and MMTS) illustrate how such studies can yield valuable information about steroid receptors that is not easily obtained by other methods. Additional reagents promise to afford even more useful data. Thus this chemical approach complements the methods of classical protein sequencing, of molecular biology, and of X-ray crystallography that are currently being used to understand receptor structure and function.

## Acknowledgements

We thank Brenda Briscoe for assistance with the typing of this paper.

## Abbreviations

Dex, dexamethasone, 9-fluoro-11$\beta$,17,21-trihydroxy-16$\alpha$-methyl-pregna-1,4-diene-3,20-dione; Dex-Mes, dexamethasone 21-mesylate; MMTS, methyl methanethiolsulfonate; HTC cells, hepatoma tissue culture cells.

## References

Arriza, JL, Weinberger, C, Cerelli, G, Glaser, TM, Handelin, BL, Housman, DE, and Evans, RM (1987) Science, 237, 268-275

Baxter, JD, and Tomkins, GM (1971) Proc. Natl. Acad. Sci., USA, 68, 932-937

Bodwell, JE, Holbrook, NJ, and Munck,A (1984a) Biochem., 23, 1392-1398

Bodwell, JE, Holbrook, NJ, and Munck,A (1984b) Biochem., 23, 4237-4242

Brocklehurst, K (1979) Int. J. Biochem., 10, 259-274

Brown, JR, (1977) In Albumin Structure, Function and Uses (VM Rosenoer, M Oratz, and MA Rothschild, eds.), Pergamon Press,New York, 27-51

Carlstedt-Duke, J, Stromstedt, P-E, Persson, B, Cederlund, E, Gustafsson, J-A, and Jornvall, H (1988) J. Biol. Chem., 263, 6842-6846

Danielsen, M, Northrop, JP, and Ringold, GM (1986) EMBO J., 5, 2513-2522

DiSorbo, DM, Phelps, DS, and Litwack, G (1980) Endocrinol., 106, 922-929

Eisen, HJ, Schleenbaker, RE, and Simons, SS Jr. (1981) J. Biol. Chem., 256, 12920-12925

Formstecher, P, Dumur, V, Idziorek, T, Danze, P-M, Sablonniere, B and Dautrevaux,M (1984) Biochim. Biophys. Acta, 802, 306-313

Gametchu, B, and Harrison, RW (1984) Endocrinol., 114, 274-279

Garner, CD, and Bristow, S (1985) in Molybdenum Enzymes (T.G. Spiro, ed.) Wiley, New York, pp 343-410

Giguere, V, Hollenberg, SM, Rosenfeld, MG, and Evans, RM (1986) Cell, 46, 645-652

Godowski, PJ, Rusconi, S, Miesfeld, R, and Yamamoto, KR (1987) Nature, 325, 365-368

Harrison, RW, Woodward, C, and Thompson, E (1983) Biochim. Biophys. Acta, 759,1-6

Hollenberg, SM, Weinberger, C, Ong, ES, Cerelli, G, Oro, A, Lebo, R, Thompson, EB, Rosenfeld, MG, and Evans, RM (1985) Nature, 318, 635-641

Kasper, CB (1975) Mol. Biol. Biochem. Biophys., 8, 114-161

Kaul, BB, Enemark, JH, Merbs, SL, and Spence, JT (1985) J. Amer. Chem. Soc., 107, 2885-2891

Koblinsky, M, Beato, M, Kalimi, M, and Feigelson, P (1972) J. Biol. Chem., 247, 7897-7904

Kumar, V, Green, S, Staub, A, and Chambon, P (1986) EMBO J., 5, 2231-2236

Mendel, DB, Bodwell, JE, and Munck, A (1986) Nature, 324, 478-480

Mendel, DB, Bodwell, JE, Smith, LI, and Munck, A (1987) in Steroid and Sterol Hormone Action (T.C. Spelsberg and R. Kumar, eds.) Martinus Nijhoff, Boston, pp 175-193

Mercier, L, Miller, PA, and Simons, SS Jr. (1986) J. Steroid Biochem., 25, 11-20

Miesfeld, R, Rusconi, S, Godowski, PJ, Maler, BA, Okret, S, Wikestrom, A-C Gustafason, J-A, and Yamamoto, KR (1986) Cell, 46, 389-399

Miller, NR, and Simons, SS Jr. (1988b) J. Biol. Chem., in press

Miller, PA, and Simons, SS Jr. (1988a) Endocrinol., 122, 2990-2998

Miller, PA, Ostrowski, MC, Hager, GL, and Simons, SS Jr. (1984)

Biochem., 23, 6883-6889

Naray, A, and Bathori, G (1982) J. Steroid Biochem.,16, 199-205

O'Brien, JM, Thanassi, and Cidlowski, JA (1980) Biochem. Biophys. Res. Commun., 92, 155-162

Okret, S, Wikstrom, A-C, Wrange, O, Andersson, B, and Gustafsson, J-A (1984) Proc. Natl. Acad. Sci., USA, 81, 1609-1613

Pons, M, Robinson, TEJ, Mercier, L, Thompson, EB, and Simons, SS Jr. (1985) J. Steroid Biochem.,23, 267-273

Rees, AM, and Bell, PA (1975) Biochim. Biophys. Acta, 411, 121-132

Reichman, ME, Foster, CM, Eisen, LP, Eisen, HJ, Torain, BF, and Simons, SS Jr (1984) Biochem., 23, 5376-5388

Rudikoff, S, and Pumphrey, JG, (1986) Proc. Natl. Acad. Sci., USA, 83, 7875-7878

Rusconi, S, and Yamamoto, KR (1987) EMBO J., 6, 1309-1315

Sanchez, ER, Toft, DO, Schlesinger, MJ, and Pratt, WB (1985) J. Biol. Chem., 260, 12398-12401

Sanchez, ER, Meshinchi, S, Tienrungroj, W, Schlesinger, MJ, Toft, DO, and Pratt, WB (1987a) J. Biol. Chem., 262, 6986-6991

Sanchez, ER, Tienrungroj, W, Meshinchi, S, Bresnick, EH, and Pratt, WB (1987b) in Steroid and Sterol Hormone Action (T.C. Spelsberg and R. Kumar, eds.) Martinus Nijhoff, Boston, pp 195-211

Simons, SS Jr. (1987) J. Biol. Chem., 262, 9669-9676

Simons, SS Jr. (1988) in Affinity Labeling in Steroid and Thyroid Hormone Research (H. Gronemeyer, ed.) Ellis Horwood, LTD, Chichester, England, pp 28-54 and 109-143

Simons, SS Jr., and Miller, PA (1984) Biochem., 23, 6876-6882

Simons, SS Jr., and Miller, PA (1986) J. Steroid Biochem., 24, 25-32

Simons, SS Jr., and Thompson, EB (1981) Proc. Natl. Acad. Sci., USA, 78, 3541-3545

Simons, SS Jr., and Thompson, EB (1982) Biochemical Actions of Hormones, 9, 221-254

Simons, SS Jr., Pons, M, and Johnson, DF (1980) J. Org. Chem., 45, 3084-3088

Simons, SS Jr., Schleenbaker, RE and Eisen, HJ (1983) J. Biol. Chem., 258, 2229-2238

Simons, SS Jr., Pumphrey, JG, Rudikoff, S, and Eisen, HJ (1987) J. Biol. Chem., 262, 9676-9680

Simons, SS Jr., Mercier, L, Miller, NR, Miller, PA, Oshima, H, Sistare, FD, Thompson, EB, Wasner, G, and Yen, PM (1988a) Cancer Research, in press

Simons, SS Jr., Miller, PA, Wasner, G, Miller, NR, and Mercier, L (1988b) J. Steroid Biochem., 31, 1-7

Sistare, FD, Hager, GL, and Simons, SS Jr. (1987) Mol. Endo., 1, 648-658

Smith, DJ, Maggio, ET, and Kenyon, GL (1975) Biochem., 14, 766-771

Smith, Ll, Bodwell, JE, Mendel, DB, Ciardelli, T, North, WG, and Munck, A

(1988) Biochem., <u>27</u>, 3747-3753

Tai, P-KK, Maeda, Y, Nakao, K, Wakim, NG, Duhring, JL, and Faber, LE
(1986) Biochem., <u>25</u>, 5269-5275

Wrange, O, Okret, S, Radojcic, M, Carlstedt-Duke, J, and Gustafsson, J-A
(1984) J. Biol. Chem., <u>259</u>, 4534-4541

DISCUSSION OF THE PAPER PRESENTED BY S.S. SIMONS

LIAO: Is there any evidence that cys 656 is located at the opening end of the Dex binding cavity as you have shown in your diagram? If there is, one may know more about how the steroid gets into the cavity.

SIMONS: I agree that such information would be very helpful in understanding the process of steroid binding to the receptor. Four indirect lines of evidence do suggest that Cys-656 is near the opening of the steroid binding cavity. First, several people (e.g., Failla et al., Lustenberger et al., Grandics et al., and Govindan and Gronemeyer) have shown that steroids that are covalently attached by the C-21 position to an affinity column matrix are very effective in binding glucocorticoid receptors. Second, charged, hydrophilic groups at the C-21 position of glucocorticoids do not appear to affect dramatically the binding affinity to the receptor if there is a reasonable distance between the C-21 carbon atom and the charged group (see Pons et al., J. Steroid Biochem., 23:267-273 [1983] and references therein). Third, Wolff et al., have calculated that hydrophobic binding accounts for most of the binding of steroids to glucocorticoid receptors and that the steroid is enveloped on both sides by the receptor protein. The C-17 dihydroxy acetone side chain, which is found on most glucocorticoids and which includes the C-21 position, is the most hydrophilic portion of the steroid and thus probably would not be in the interior of the binding cavity. Finally, the reactive C-21 position of dexamethasone-mesylate that causes the covalent labeling of the receptor is on this hydrophilic extremity of the steroid; thus, it is possible that the bulk of the steroid binding cavity is somewhat distant from the C-21 position and the affinity labeled Cys-656 of the receptor. Collectively,

these data suggest that the C-21 position of most glucocorticoids
and the affinity labeled Cys-656 are relatively exposed to solvent
and at an "opening" of the steroid binding cavity.

THOMPSON: Why does MMTS lower the affinity of dex mesylate whereas
it does not do so for dexamethasone, a compound with very similar
structure?

SIMONS: MMTS pretreatment of receptors lowers the absolute affinity
of dexamethasone by a factor of about 5. The answer to the question
of why MMTS pretreatment lowers the affinity of receptors for
dexamethasone-mesylate more than for dexamethasone has two parts.
First, without knowing the precise geometry of the steroid binding
cavity and the placement of the MMTS modified thiols, it is difficult
to present a totally analytical explanation. Our attempts to obtain
some information regarding the geometry of the binding site have been
unsuccessful because we have not yet been able to discern a consistent
relationship between steroid structure and affinity for receptors $\pm$
MMTS modification. It should be noted that the affinity  of dexamethasone-oxetanone (another compound with a structure very similar
to that of dexamethasone) for receptors is also reduced by MMTS
pretreatment more than for dexamethasone. On the other hand, MMTS
pretreatment does not change the absolute affinity of receptors for
Deacylcortivazol or for RU 38,486 (two steroids with structures very
different from dexamethasone) so that the affinity of these steroids,
relative to dexamethasone, is actually increased by MMTS.

Second, the statement that the affinity of dexamethasone-mesylate for
MMTS modified receptors is reduced is based on our assumption that the
effects of MMTS pretreatment are the same for steroids of similar
structures, i.e. dexamethasone-mesylate, dexamethasone-oxetanone,
and cortisol. This approach is necessary because dexamethasone-

mesylate binding to untreated receptors is a combination of non-covalent and covalent binding and we do not know the contribution of these two processes to the observed $K_a$. Unfortunately, as we saw above, relatively minor changes in steroid structure can have major effects on steroid binding affinity. Therefore this type of comparison may not work. Strictly speaking, all we can say is that MMTS modified glucocorticoid receptors have an affinity for dexamethasone-mesylate that is the same as that for the structurally similar steroids dexamethasone-oxetanone and cortisol, each of which has a 5-fold reduced affinity.

DISCUSSANTS: S. LIAO, S.S. SIMONS AND E.B. THOMPSON.

# REGULATION OF GLUCOCORTICOID RECEPTOR PROTEIN AND GENE EXPRESSION BY GLUCOCORTICOIDS

Yan Min Wang[^], Kerry Burnstein[*], Corinne Silva[#], Deborah Bellingham[#], Douglas Tully[#], Jorge Simental[+], Christine Jewell[+], and John A. Cidlowski[*#+] Curriculum in Neurobiology[^], Cancer Research Center[*], Departments of Biochemistry[#] and Physiology[+], University of North Carolina at Chapel Hill, Chapel Hill, North Carolina, 27599

## Introduction

Glucocorticoids, members of the highly conserved steroid hormone family in evolution, exert numerous physiological effects on the developmental and adaptational processes of eukaryotic organisms. In general, free circulating glucocorticoids released from the adrenal cortex in response to adrenocorticotropic hormone (ACTH) from the pituitary enter target cells by a passive diffusion process conferred by the high lipid solubility common to all steroids. Although there have been some reports on membrane effects of steroid hormones, actions of steroid hormones appear to be largely mediated by their intracellular receptors. These receptors act much like ligand-dependent transcription factors that regulate gene expression. The magnitude of steroid hormone responses, therefore, is determined not only by the hormone concentration but also by cellular levels of functional receptor proteins. Proposed models for the structure of the glucocorticoid receptor (GR) have the common feature that the receptor exists as an oligomeric protein complex that consists of one or more subunits complexed with non-steroid-binding proteins such as the heat shock protein 90 (hsp90). Dissociation of the oligomeric complex takes place following binding of the hormone and permits translocation of the steroid-receptor complex into the nucleus where it associates with enhancer-like glucocorticoid response elements (GREs) in the genome. This could either induce the expression of normally silent genes or change the transcription of constitutively expressed housekeeping genes (such as the GR gene itself) by up- or down-regulation. Thus to fully understand the mechanism underlying glucocorticoid action, the study of GR autoregulation becomes an imperative issue.

The study of GR regulation has been greatly facilitated by the availability of several highly useful tools. For one, the ability to synthesize tritium($^3$H)-labelled steroid hormones with high specific activity has aided receptor binding assays tremendously and precipitated significant information about GR regulation both in cell culture and in whole animal studies. Partial purification of GR made the *in vitro* structure-function studies of GR feasible in addition to generating material for the production of antibodies. For instance, it is well accepted by now that the monomeric subunit of GR consists of three functional domains: a regulatory domain at the amino terminal that is highly immunogenic; a small DNA binding domain in the middle; and a steroid binding domain on the carboxyl end (Gustafsson et al. 1987). Although not much information about GR down-regulation has been gathered yet using the antibodies, they are potentially powerful tools for studying GR regulation at the protein level. Affinity labels, such as dexamethasone mesylate (Simons and Thompson 1981), also represent potentially useful tools for studying GR level changes even though they have not been widely employed. Perhaps the most important discovery which has led to renewed interest in the regulation of GR has been the cloning of the complementary DNA for GR (Miesfeld et al. 1984, Hollenberg et al. 1985), which has been used for Northern blot analysis, slot or dot blot hybridization, transcription assays, transfection studies, and mutagenesis studies. The discovery of GRE (Scheidereit et al. 1983, Payvar et al. 1983), on the other hand, presented the opportunity to directly study the mechanisms underlying glucocorticoid regulation of gene expression, including that of its own gene, in contrast to the conventional use of physiological indexes (for more comprehensive reviews about properties of GR, see Gustafsson et al. 1987 and Rousseau 1984).

In this manuscript, we present selected data from studies performed both in our own and other laboratories using these tools which may provide new insights concerning the regulation of GR at both protein and mRNA levels.

<u>Retrospect</u>

Down-Regulation of Steroid Binding in Cell Culture

One of the most remarkable features of GR is its ubiquity, for it can be found in virtually every type of cell in the body. It is involved in a wide range of physiological functions and is also susceptible to regulation. There are conceivably two ways to regulate GR, either up or down. Presumably, up-regulation enhances steroid responsiveness while down-regulation desensitizes a cell to steroid. Most of the studies carried out so far deal with down-regulation of GR, which has more direct clinical relevance since glucocorticoid excess is a fairly common medical condition and synthetic glucocorticoids are widely used for treating various types of ailments.

In 1981, two independent reports demonstrated that glucocorticoid binding decreases in both the cytosol and the nucleus of two lines of cultured tumor cells (mouse pituitary AtT-20/D-1 and human cervix epithelium HeLa S3) after they have been incubated with glucocorticoids (Svec and Rudis 1981; Cidlowski and Cidlowski 1981). This reduction in glucocorticoid binding was interpreted to be down-regulation of the GR protein. An example of this effect is seen in Figure 1. HeLa cells were pretreated for 24 hours with $10^{-7}$M dexamethasone (DEX) and nuclear binding was quantitated using a whole cell assay. The cells were

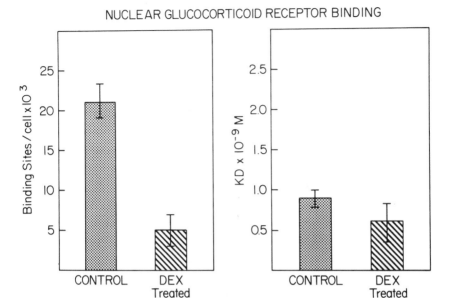

Fig. 1. Scatchard analysis of nuclear glucocorticoid receptor binding in Hela S3 cells after they were incubated with $10^{-7}$M dexamethasone (DEX) for 24 hours at 37C. Binding was measured after the steroid had been removed with extensive washing.

washed extensively to remove all detectable steroid prior to measurement of binding (Cidlowski and Cidlowski 1981). Subsequently, cell nuclei were isolated and the specific activity counted. The data were plotted by the method of Scatchard and the $B_{max}$ and $K_d$ values obtained. As can be seen, DEX treatment of these cells resulted in a marked reduction of steroid binding sites per cell nucleus (approximately 5000/cell nucleus), as compared to about 20,000 sites in the control cell. On the right of the figure, $K_d$ values are compared for the control and treated groups. The $K_d$ values are very comparable for the two groups, indicating that no unlabelled steroid is present in these cells. Based on the observation that virtually all of the steroid used in pretreatment could be removed from these cultures, we and Svec and Rudis proposed that the GR protein was indeed down-regulated.

In order to determine if this GR down-regulation phenomenon is a receptor-dependent process, steroid specificity studies were performed. Figure 2 shows binding sites per nucleus for control and dexamethasone, progesterone, estradiol and 5 -dihydrotestosterone pretreated cell groups. Apart from the partially effective progesterone, which has known antiglucocorticoid properties, only

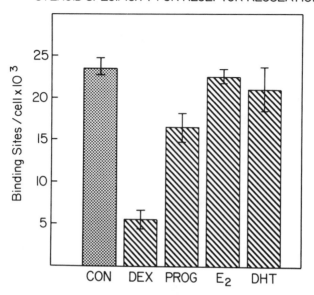

STEROID SPECIFICITY FOR RECEPTOR REGULATION

Fig. 2. Steroid specificity test for glucocorticoid receptor down-regulation by dexamethasone (DEX). Assay conditions were the same as those used for Fig. 1, except that unlabeled steroids were used. Con = control; DHT = 5$\alpha$-dihydrotestosterone; E$_2$ = estradiol; Prog = progesterone.

pretreatment with DEX produced a drastic decrease in glucocorticoid binding sites in the cell. These data are also supported by a similar line of experiments by Svec and Rudis (1981). In their studies, pretreatment with progesterone and aldosterone, which are known to interact with GR, and DEX and corticosterone were able to cause an appreciable reduction in specific binding. After the potent glucocorticoid antagonist RU486 became available, Rajpert et al. (1987) suggested that down-regulation of GR is probably an agonist-mediated phenomenon that requires receptor transformation. Although RU486-bound GR interacted with GRE-containing DNA fragments *in vitro*, it failed to effectively down-regulate steroid binding at least in IM-9 cells. Whether this is due to a genuine antagonistic property of the drug or results from inadequate cellular access remains unclear. Considered together, the above experiments indicate that GR down-regulation, as measured by steroid binding, is a receptor-dependent process caused specifically by glucocorticoids. In order to learn more about the process of down-regulation, we initiated an ongoing experiment to elucidate the long-term kinetics of GR down-regulation in the HeLa S3 system. We found that HeLa only cells which have

KINETICS OF GLUCOCORTICOID INDUCED DOWN REGULATION

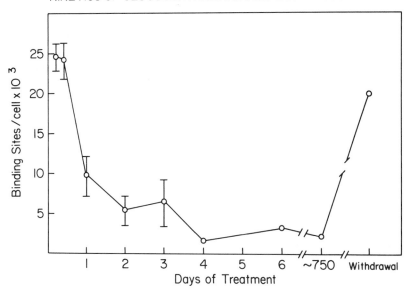

Fig. 3. Long-term kinetics of dexamethasone induced down-regulation of glucocorticoid receptors and their recovery after hormone withdrawal.

been maintained in the presence of DEX for a period of over two years have GR levels which, remarkably, are as low as (if not lower than) that in the initially treated cells (Fig. 3). Upon removal of the hormone, however, the GR level gradually climbs back to near the original level. This suggests an active regulatory mechanism present in the cell as opposed to selective survival of a subclone or an artifact caused by the steroid. Later work by Seigler and Svec (1984) further revealed that after a short treatment, and a similar steroid removal regime, the recovered form of GR in AtT-20 cell line is indeed identical to the original form. Our unpublished work supports this observation. Therefore, it appears that continuing hormone exposure is manditory for down-regulation of steroid binding to persist.

The ability of glucocorticoids to down-regulate their own receptors as measured by steroid binding has since been confirmed in several other cell culture systems. GR down-regulation induced by a variety of glucocorticoids was also found to occur in several mouse T-lymphosarcoma subclones derived from the parent cell line WEHI7 (Danielsen and Stallcup 1984), in a primary culture of human T lymphocytes maintained with T cell growth factor (Lacroix et al. 1984), in a rat pituitary cell line GH1 (McIntyre and Samuels 1985), and in the human lymphoblastoid IM-9 cells (Rajpert et al. 1987). These results help to confirm GR down-regulation as a widespread phenomenon.

Down-regulation of Steroid Binding in Animals

Although the above mentioned studies using cultured cells have the advantages of cell homogeneity, feasibility of cell synchrony and discrete control of the chemical milieu, they risk the potential pitfalls of cell culture and can not substitute for whole animal studies that carry direct physiological relevance. Down-regulation of GR was observed in the rat hippocampus after chronic treatment with corticosterone, the naturally occurring glucocorticoid in rodents (Tornello et al. 1982). Later, more extensive work revealed that the same phenomenon occurs after the animal has been placed under repeated stress, which is known to cause a high level of corticosterone secretion (Sapolsky et al. 1984). The phenomenon of GR down-regulation has also been observed clinically, first by Schlechte et al. (1982), then by Shipman et al. (1983). Lymphocytes were collected from peripheral blood of patients treated with glucocorticoids for varying lengths of time and steroid binding assays performed. In both cases, GR down-regulation was observed and was followed by a gradual return to base-line level upon steroid removal. Together, these studies serve to establish GR down-regulation as a genuine physiological process.

Physiological Consequences of Down-Regulation

The physiological relevance of GR down-regulation is also reflected in changes in gene expression of specific proteins known to be regulated by glucocorticoids. For example, alkaline phosphatase (AP) is known to be specifically regulated by glucocorticoids in the HeLa S3 cells (Littlefield et al. 1980). We have examined the influence of chronic DEX treatment that promotes down-regulation, on the stimulation of AP activity in HeLa S3 cells. Figure 4 shows a biphasic change in AP activity after extended exposure to increasing amounts of DEX. At a concentration of DEX close to the dissociation constant ($K_d$) of the drug ($10^{-9}$M), enzyme activity is increased compared to the control (untreated) level and fluctuates around 15 nkatels. When the DEX concentration is increased to 10 x $K_d$ ($10^{-8}$M), the enzyme activity climbs to a peak value. This activity is followed by a steep fall which is again succeeded by a smaller increase. With increasing DEX concentration ($10^{-7}$M then $10^{-6}$M) and increasing length of exposure time (up to 45 days), the activity stayed well below control level with only one incidental overshoot. This attenuation of steroid responsiveness could be correlated with the GR down-regulation seen in Figure 3. Thus in our studies, conditions which cause GR down-regulation lead to reduced

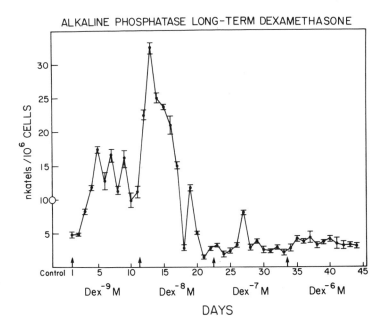

Fig. 4. Effects of chronic treatment with incremental dexamethasone concentrations on alkaline phosphatase activity in HeLa S3 cells. For methodology, see Littlefield et al. (1980).

ability of DEX to induce AP activity. Another example of gene expression possibly related to GR down-regulation is the aromatase gene present in cultured human foreskin fibroblasts. Recently, Berkovitz et al. (1988) demonstrated that DEX treatment caused both GR down-regulation and a biphasic response of aromatase activity much like what we observed with AP in HeLa S3 cells. Down-regulation, however, ocurred before peak enzyme activity was achieved. Thus it is impossible to directly correlate these phenomena. The important point coming from both the AP and aromatase studies is that GR down-regulation is associated with dampened glucocorticoid-dependent gene expression. Whether other mechanisms are involved in this process remains unclear.

Limitation of the Steroid Binding Approach

One caveat to the above-mentioned studies including our own work on GR down-regulation is that only steroid binding is being measured. Thus these studies do not address mechanistic questions of receptor down-regulation. This limitation set by the technology of the time is now being overcome and will be discussed in the next section.

Table 1. Potential Mechanisms Involved in Glucocorticoid
Receptor Down-Regulation

1. Increased Receptor Degradation
   A. general protein degradation
   B. ubiquitin-dependent degradation

2. Decreased Receptor Synthesis

   A. altered mRNA stability
   B. reduced transcription initiation
   C. transcription attenuation

3. Receptor Inactivation/Modification

   A. non-steroid-binding forms
   B. non-DNA-binding receptors

Possible Mechanisms of GR Down-regulation

The cause of GR down-regulation could be multifold. Table 1 lists several of the
proposed potential mechanisms of GR down-regulation. For instance,
glucocorticoids could down-regulate GR level by inducing or increasing the
expression of a specific protease that in turn speeds up the degradation of GR, thus
decreasing its half-life. Alternatively, binding of the hormone to GR and its
subsequent activation (also termed transformation) could lead to faster degradation
of the protein. The aforementioned paper by McIntyre and Samuels (1985)
supports the view that triamcinolone acetonide treatment of cells decreases the half-
life of the transformed nuclear form of GR in a protein synthesis independent
manner. Another possible explanation of GR down-regulation is that ubiquitination
of GR could somehow occur, which could in turn targets the receptor for
degradation. As of now, however, there is no experimental evidence to substantiate
this hypothesis. Instead of increased GR degradation taking place, the synthesis of
GR could be decreasing. Such a situation could occur if glucocorticoids induce or
increase the transcription of a labile RNase which acts specifically on GR mRNA.
The recent report by Yen et al. (1988) concerning b-tubulin mRNA autoregulation
by an amino-terminal-dependent nuclease provides an elegant example of what
could be involved in the process. Another possibility leading to decreased synthesis
of GR mRNA is that the steroid-receptor complex could directly affect the initiation
or attenuate the transcription of the GR gene (i.e. reducing the rate of either

initiation or elongation). A further discussion of this possibility will follow. Prompted by some recent data generated by Corinne Silva in our laboratory as well as data from other laboratories (Mendel et al. 1986), we are led to believe that decreased steroid-binding of GR observed during down-regulation is partially due to generation of non-steroid-binding forms of GR during receptor recycling. This will also be discussed later. Cellular production of a non-DNA-binding form of GR, on the other hand, could also give rise to GR down-regulation if the regulatory step involves positive GR interaction with GREs in the GR gene. As of now, however, there is really insufficient evidence to formally prove any of these possible mechanisms. It is likely that different mechanisms are at work in different tissues and that a combination of these possibilities coexist.

Steady State Glucocorticoid Receptor mRNA

With the ever-increasing application of molecular biological techniques, the possibilities mentioned above are now being examined. In 1986, Okret et al. published the first report of GR mRNA down-regulation (50-95%) by DEX in both a rat hepatoma cell line (H4IIE) and in rat livers by using Northern blot analysis. They demonstrated that this phenomenon is independent of protein synthesis by using the translation inhibitor cycloheximide. More recently, Kalinyak et al. (1987) performed a more comprehensive study of the DEX effect on GR mRNA in various tissues of the rat. First of all, a drastic difference due to adrenalectomy was seen on Northern and slot blots of GR mRNA. Whereas brain and kidney showed 40% and 80% increases respectively two weeks after surgery, the changes in liver and lung were negligible. After six hours of DEX treatment, however, a fairly uniform decrease in GR mRNA ranging from 40% to 60% was seen in all tissues examined, including adrenal, spleen, heart and testes in addition to the above mentioned four tissues. This comparative approach helped reveal potential tissue-specific mechanisms of GR message autoregulation. Nevertheless, use of the whole brain as a tissue is inappropriate, and examination of more discrete regions is needed. With similar goals in mind, our laboratory has examined DEX effects on GR mRNA in the original HeLa S3 cell line, where GR down-regulation was first observed, and in the rat hippocampus. Figure 5 shows on the left panel a time course of DEX treatment with poly $(A)^+$ RNA isolated from the hippocampi of adrenalectomized rats. GR mRNA level was noticeably decreased within two hours and obvious reduction in steady state levels occurred four and five hours after hormone administration. The right panel of Figure 5 shows a Northern blot containing poly $(A)^+$ RNA isolated from two separate groups of HeLa $S_3$ cells that were either

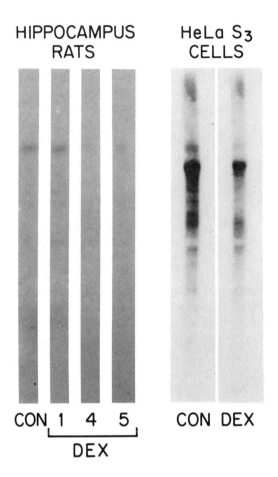

Fig. 5. Northern blots containing poly (A)$^+$ mRNA isolated from rat hippocampus and HeLa S$_3$ cells. Adrenalectomized rats were treated with either vehicle (control) or a one milligram dose of dexamethasone (DEX) for 1-5 hours and sacrificed at four time points (DEX: 1, 4 and 5 hours). HeLa S$_3$ cells were either left untreated (CON lane) or exposed to DEX for 24 hours (DEX lane). Con = control; DEX = dexamethasone.

subjected to DEX treatment for 24 hours (right) or left untreated (left). As can be seen, the GR mRNA levels in the DEX-treated cells are considerably lower than in the control cells. Thus, GR down-regulation can be detected not only by steroid binding, as was presented, but also at the mRNA level.

Glucocorticoid Receptor mRNA Transcription and Stability

Although the above reports collectively suggest that down-regulation of GR occurs at the mRNA level rather than or in addition to the protein level, it is still not clear whether glucocorticoids down-regulate GR by directly affecting the transcription or by decreasing the message half-life via an RNase. A recent report by Rosewicz et al. (1988) supports the first possibility. It was shown with nuclear run-off assays that in the human lymphoblastoid cells (IM-9) where DEX treatment clearly results in GR mRNA decrease, the transcription rate of the GR gene declined to $50\pm6\%$ of the control after DEX. It was also shown that the GR mRNA half-life, which was predetermined to be about two hours in this cell line, remains unaffected by DEX. One potential criticism of this mRNA half-life experiment is that the measurements were taken after down-regulation has already taken place instead of while it is happening. Since this is the only report of a direct glucocorticoid effect on GR gene transcription, it still remains to be seen whether the same situation exists in other cell types, especially considering the existence of the diverse tissue-specific regulatory mechanisms reported so far.

Prospect

Down-regulation of Glucocorticoid mRNA in Transfected Cells

A possible explanation for the attenuation of the rate of transcription of GR mRNA observed by Rosewicz et al. (1988) could relate to a direct interaction of activated GR with its own gene. Douglas Tully in our laboratory first made the observation that the original GRE hexanucleotide core sequence (TGTTCT) discovered by Scheidereit et al. (1983) and Payvar et al. (1983) can be found in multiple copies in the cDNA of hGR. More recently, using the partially symmetrical consensus sequence (GGTACANNNTGTTCT) compiled by Evans (1988) as a starting point, Jorge Simental in our laboratory performed a computer search for this sequence in the hGR cDNA. Figure 6 shows the result of this search. Five sequences were found within the hGR coding region on the non-coding strand. These sequences fit with either part or the whole of the consensus sequence presented above it, given two positions of ambiguity on the left arm of the sequence. Is it possible that these potentially functional GREs are actually the cause of or at least involved in GR down-regulation? Kerry Burnstein in our laboratory has been pursuing this question. The human glucocorticoid receptor (hGR) cDNA driven by the glucocorticoid-independent RSV promoter was transiently transfected into the SV40 transformed African green monkey kidney cell line (COS-1). Down-regulation of GR mRNA was observed after the cells were treated with $10^{-7}$M DEX for 15-17 hours. When the cells were transfected with an RSV-driven chloramphenicol

# HUMAN GLUCOCORTICOID RECEPTOR GENE

## Potential Glucocorticoid Regulatory Elements (GRE's)

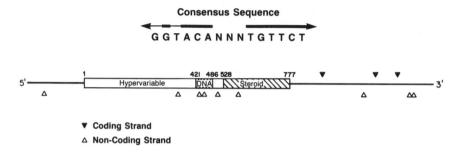

Fig. 6. Potentially functional glucocorticoid regulatory elements (GREs) found within the human glucocorticoid receptor gene. The consensus sequence used was taken from Evans (1988).

acetyltransferase (CAT) cDNA, however, DEX failed to induce down-regulation of this enzyme as measured by CAT activity assays. These experiments imply that there is sufficient information encoded in the hGR cDNA to permit GR down-regulation. Another observation made by Deborah Bellingham in our laboratory also supports this view. While studying regulation of a stably transfected glucocorticoid-dependent gene in a cell line, she observed that, as compared to the GR mRNA level in untreated cells, the expression of the cotransfected RSV-driven hGR cDNA decreased substantially after the cells were maintained on a low level (5 nM) of DEX for 2-3 months. These two lines of evidence together suggest as a potential mechanism of GR down-regulation that binding of activated GR to GREs found within the coding region of the hGR gene could obstruct elongation of the GR mRNA by RNA polymerase II.

Non-steroid Binding Form of Glucocorticoid Receptor

Experiments performed by Corinne Silva in our laboratory point to an additional mechanism for GR regulation. By using [$^3$H]-dexamethasone mesylate affinity labelling as well as Western blotting with a polyclonal antibody (Wilson et al. in press), she found that although the GR levels dropped below the level of detection by [$^3$H]-dexamethasone mesylate labelling after chronic DEX treatment, the immunoreactive level of GR as seen by Western blotting remained comparable to control levels. This suggests that there exists in the apparently down-regulated cells a non-steroid-binding form of GR that is fully immunoreactive. The origin of this non-steroid-binding form remains a point of speculation. One explanation is that this represents an intermediate form of the receptor undergoing the process of recycling. Another possibility is that there are actually two alleles encoding two slightly different forms of GR in HeLa $S_3$ cells. An example of such heterozygous expression can be found in the wild type mouse thymoma S49 cells (Westphal et al. 1984; Northrop et al. 1985). We are tending, however, toward the intermediate form idea, for Mendel et al. (1986) have revealed in their studies a non-steroid-binding nuclear form of GR by using a ATP-depleting regime. It is conceivable that an overwhelming dose of hormone exposure, incurred during acute stress for example, may lead to considerable energy depletion in the hormone responsive target cell, a so-called "fatigue" state. Such a state will produce a lag in the cell's ability to recycle the receptor from a non-steroid-binding intranuclear form back to the cytosolic steroid binding form, which is presumably an ATP-dependent process (Munck and Brinck-Johnsen 1968). Equally possible, however, is the heterozygous expression theory just mentioned above, wherein only the functional steroid binding form of receptor is down regulated by ligand. It is worth pointing out, however, that this situation may be unique to transformed cells since they provide the only documented cases where more immunoreactive GR than steroid-binding GR were found (S49 and HeLa $S_3$ cell lines). If this were the case in normal organisms, the frequency of occurrence of a homozygote carrying the non-steroid-binding GR allele would be too high and we would expect to see much higher clinical occurrences of glucocorticoid insensitivity, which may even be fatal in an early stage of life. Since most of the data just mentioned in this section are either preliminary or from isolated reports, they remain good prospects open to future investigation.


Concluding Remarks

A variety of glucocorticoids are capable of down-regulating their own receptor concentration in the cell. This down-regulation has been observed both at the steroid binding and at the mRNA levels by employing labelled steroids, nucleic acid probes, affinity ligands and antibodies. The mechanisms underlying this down-regulation remain unclear. Some experimental evidence points to transcription attenuation, and/or GREs within the GR coding region as potential mechanisms. Evidence also exists suggesting generation of a non-steroid-binding form of GR during increased demand for recycling. Yet another possibility may be the heterozygous expression of two forms of GR with only one being responsive to the hormone. Nevertheless, many questions remain. For instance, the fact that the GR protein never disappears from the cell in spite of extended exposure to high levels of glucocorticoids has yet to be addressed. Elucidation of this phenomenon may help reveal the mechanism responsible for such a stringent degree of down-regulation. Another possibility that may account for GR mRNA down-regulation in some cases is a simple dilution effect. Although only countable numbers of genes have so far been established as being under steroid regulation, general stimulatory effects on RNA synthesis may proportionally dilute out GR mRNA in the cytoplasm. Alternatively, receptor levels in cells could be controlled by DNA-independent mechanisms such as Gill and Ptashne (1988) recently proposed for the yeast transcriptional activator GAL4. Given the complexity of gene expression regulation in eukaryotic organisms, the list of possibilities could be endless. The study of GR regulation, therefore, remains an open field and awaits further examination.

*Acknowledgments.* Research performed in our laboratory has been supported by the following grants from NIH AM-32460, AM-32459 and AM-32078. We also wish to thank Robert and Toby Schwartzman for generous help during the preparation of this manuscript.

References

Berkovitz GD, Carter KM, Migeon CJ, Brown TR (1988) J Clin Endocrinol Metab 66:1029-1036

Cidlowski JA, Cidlowski NB (1981) Endocrinology 109:1975-1982

Danielsen M, Stallcup MR (1984) Mol Cell Biol 4:449-453

Evans RM (1988) Science 240:889-895

Gustafsson J, Carlstedt-Duke J, Poellinger L, Okret S, Wikstrom A, Bronnegard M, Gilliner M, Dong Y, Fuxe K, Cintra A, Harfstrand A, Agnati L (1987) Endocrine Reviews 8:185-234

Gill G, Ptashne M (1988) Nature 334:721-724

Hollenberg SM, Weinberger C, Ong ES, Cerelli G, Oro A, Lebo R, Thompson EB, Rosenfeld MG, Evans RM (1985) Nature 318:635-641

Kalinyak JE, Dorin RI, Hoffman AR, Perlman AJ (1987) J Biol Chem 262:10441-10444

Lacroix A, Bonnard GD, Lippman ME (1984) J Steroid Biochem 21:73-80

Littlefield BA, Cidlowski NB, Cidlowski JA (1980) Arch Biochem Biophys 201:174-184

McIntyre WR, Samuels HH (1985) J Biol Chem 260:418-427

Mendel DB, Bodwell JE, Munck A (1986) Nature 324:478-480

Miesfeld R, Okret S, Wikstrom A-C, Wrange O, Gustafsson J-A, Yamamoto K, (1984) Nature 312:779-781

Munck A, Brinck-Johnsen T (1968) J Biol Chem 243:5556-5565

Northrop JP, Gametchu B, Harrison RW, Ringold GM (1985) J Biol Chem 260:6398-6403

Okret S, Poellinger L, Dong Y, Gustafsson J-A (1986) Proc Natl Acad Sci USA 83:5899-5903

Payvar FP, DeFranco D, Firestone GL, Edgar B, Wrange O, Okert S, Gustafsson J-A, Yamamoto KR (1983) Cell 35:381-392

Rajpert EJ, Lemaigre FP, Eliard PH, Place M, Lafontaine DA, Economidis JV, Belayew A, Martial JA, Rousseau GG (1987) J Steroid Biochem 26:513-520

Rosewicz S, McDonald AR, Maddux BA, Goldfine ID, Miesfeld RL, Logsdon CD (1988) J Biol Chem 263:2581-2584

Rousseau GG (1984) Mol Cell Endocrinol 38:1-11

Sapolsky RM, Krey LC, McEwen BS (1984) Endocrinology 114:287-292

Sapolsky RM, McEwen BS (1985) Brain Res 339:161-165

Scheidereit C, Geisse S, Westphal HM, Beato M (1983) Nature 304:749-752

Schlechte JA, Ginsberg BH, Sherman BM (1982) J Steroid Biochem 16:69-74

Seigler L, Svec F (1984) Biochim Biophy Acta 800:111-118

Shipman GF, Bloomfield CD, Gajl-Peczalska KJ, Munck AU, Smith KA (1983) Blood 61:1086-1090

Simons Jr SS, Thompson EB (1981) Proc Natl Acad Sci USA 78:3541

Svec F, Rudis M (1981) J Biol Chem 256:5984-5987

Svec F (1985) J Steroid Biochem 23:669-671

Tornello S, Orti E, Nicola FD, Rainbow TC, McEwen BS (1982) Neuroendocrinology 35:411-417

Westphal HM, Mugele K, Beato M, Gehring U (1984) EMBO J 3:1493-1498

Wilson EM, Lubahn DB, French FS, Jewell CM, Cidlowski JA (in press) Mol Endocrinol

Yen TJ, Machlin PS, Cleveland DW (1988) Nature 334:580

DISCUSSION OF THE PAPER PRESENTED BY J. CIDLOWSKI

MOUDGIL:  In human breast cancer cells, T47, progesterone-dependent phosphorylation has been implicated in the regulation of progesterone receptor.  A definite increase in phosphorylation of glucocorticoid receptor has not been demonstrated, although glucocorticoid receptor is known to be a phosphoprotein.  Have you entertained the idea that phosphorylation – dephosphorylation reactions contribute to down regulation of glucocorticoid receptor in your system?

CIDLOWSKI:  Phosphorylation/dephosphorylation may certainly be involved in down regulation, but we have not done any experiments to consider this possibility.

TCHEN:  If antibodies only bind to activated receptors, how do they bind to steroid non-binding receptors?

CIDLOWSKI:  Out antibodies recognize only "activated" glucocorticoid receptors in under native conditions.  Denatured receptors such as occur on western blots are also recognized.  Our interpretation is that the "epitope" is precluded in the native unactivated receptor.

SIMONS:  We have found that low levels of certain oxidizing agents can eliminate all steroid binding.  In view of this observation, have you looked at the cell-free binding of receptors in the down-regulated cells to see if the loss of steroid binding activity (with unchanged levels of receptor protein) is due to a change in the oxidation state of the receptor?

CIDLOWSKI:  All of our studies to date have been done in whole cells and we have not made any attempt to change the oxidation reduction state of the cells.  As for a potential role of oxidation/reduction of steroid binding, I am in agreement with your hypothesis.  Corinne Silva in my lab has shown that oxidation/reduction state influences receptor folding and this type of effect could easily account for the loss of steroid binding capacity.

SIMONS: In the Southwestern blots, you see receptor cDNA binding to steroid-free receptor. Most people see a requirement of bound steroid for glucocorticoid receptor binding to DNA. Would you please comment on this.

CIDLOWSKI: We do not see DNA binding on Southwestern blots without steroid. I believe the reason for this stems from the fact that electrophoresis of the protein probably does the same thing (removal of HSP90) as steroid does in the cell.

THOMPSON: Have you done "nuclear run off" experiments to test your Hela cells for effects of glucocorticoids on the GR gene? If so, with what result?

CIDLOWSKI: We have just begun to do run off transcription on the human glucocorticoid receptor gene in HeLa $S_3$ cells. Our results are too preliminary to comment.

THOMPSON: Please discuss your results vis a vis Miesfeld's observations of decreased transcription of the GR gene in IM9 cells after glucocorticoid treatment.

CIDLOWSKI: I think our data are quite consistent with Roger's observations. It is my hypothesis that glucocorticoid receptor association with its own gene causes transcriptional pausing which should be reflected by a decrease in run off transcription and a net reduction in steady state mRNA for hGR.

THOMPSON: Do you think that GR protein is stabilized in your Dex-treated cells? It would seem so, since the mRNA is depressed in quantity and the total GR protein is the same as in untreated cells.

CIDLOWSKI: There certainly may be some receptor stabilization in our long term Dex-treated cells. The major point of these observations, however, is that we still see mRNA being produced without apparent steroid binding capacity.  Thus, I believe down regulation reflects

a loss of steroid binding capacity and not a loss of protein.

DISCUSSANTS: V. MOUDGIL, J. CIDLOWSKI, T. TCHEN, S.S. SIMONS AND

E.B. THOMPSON.

MODULATION OF GLUCOCORTICOID-INDUCED RESPONSES BY CYCLIC AMP
IN LYMPHOID CELL LINES.


Donald J. Gruol, Maureen T. Harrigan and Suzanne Bourgeois


Introduction

Glucocorticoid hormones affect immune function, in part, through their
capacity to elicit a cytolytic response in thymocytes (Claman, 1972;
Thompson and Lippman, 1974).  The lytic process is dependent upon
continued RNA and protein synthesis, and may represent induced
transcription of one or more specific genes whose expression leads to
programmed cell death (Munck and Crabtree, 1981; Young et al., 1980).
Evidence for the existence of "lysis gene(s)" has been obtained by
cell fusion experiments (Gasson and Bourgeois 1983a; Gasson and
Bourgeois, 1983b) and by work demonstrating reactivation of cytolysis
in the steroid-resistant SAK-8 cell line (Gasson et al., 1983;
Bourgeois and Gasson, 1985).  While cell cycle arrest (Harmon et al.,
1979) and DNA degradation (Wyllie, 1980) have been found to be part of
the autolytic process, the exact nature of the lysis-mediating
function(s) remains to be discovered.  It is important to keep in
mind, however, that in the intact animal, thymocyte killing resulting
from release of adrenal steroids is often part of an overall reaction
to stress (Baxter and Rousseau, 1979).  This raises the possibility
that other stress-related signals, such as catecholamines, might also
be integrated into the lymphoid cytolytic program.  Thus, in T-cells,

second messenger systems such as cAMP, calcium and phospholipid metabolites may play a contributing role in modulating the action of glucocorticoids.

During the last several years this laboratory has been investigating the effect of cAMP regulation on glucocorticoid-induced cytolysis in the murine lymphoma WEHI-7 (W7TG or W7TB cell lines). Our studies have been aided by the fact that cAMP also elicits cytolysis in these cells (Gruol and Dalton, 1984). This behavior has allowed selection of WEHI-7 variants containing defective cyclic AMP-dependent protein kinase (cAPK) activity. The results of our work have shown that: 1) Cyclic AMP has the capacity to increase the glucocorticoid binding capacity in WEHI-7 cells (Gruol et al., 1986b). 2) Loss of cAPK activity (without loss of steroid binding) causes a decreased responsiveness to glucocorticoids (Rajah et al., 1988). 3) Steroid-resistant variants can be obtained at a much higher frequency ($>10^3$ fold) from cells containing defective cAPK activity than from wild type cells (Gruol et al., 1986a; Gruol et al., 1986b).

The results cited above suggest that cAMP may modulate the response of T-cells to glucocorticoids at more than one level. We are currently evaluating this possibility by working on several related hypotheses. First, we are asking if cAMP regulates the level of glucocorticoid receptors by affecting transcription of the receptor gene. Glucocorticoids have been shown to down-regulate the synthesis of their own receptor mRNA (Okret et al., 1986; Kalinyak et al., 1987; Rosewicz et al., 1988). Regulation by cAMP may act to reverse this process. Second, cAPK may affect covalent modification of the receptor or a related factor. The glucocorticoid receptor has been shown to be phosphorylated (Housley and Pratt, 1983; Mendel et al., 1987) and is possibly a substrate for cAPK (Singh and Moudgil, 1985). Changes in receptor modification could affect the efficiency of receptor function. Finally, we would like to know if cAMP acts in concert with glucocorticoids to regulate genes which play a role in the lytic response. There are a number of examples where glucocorticoids and cAMP act together to regulate expression of specific genes (Granner and Hargrove, 1983; Short et al., 1986). The lytic gene(s) may be another example of this class.

## Results

### Cyclic AMP Regulation of Steroid Binding Capacity

Murine lymphoma cell lines such as WEHI-7 and S49 are killed by exposure to both glucocorticoids and signals which increase their intracellular levels of cAMP (Gruol and Bourgeois, 1987). This sensitivity has provided a convenient means to select for variant cell lines resistant to both types of agents. In the case of cAMP resistance (cAMP$^r$), the majority of variants display an altered cAMP-dependent protein kinase (cAPK) activity (Coffino et al., 1976). Figure 1 illustrates two kinase phenotypes that have been observed in lymphoma cells. It shows the dose-response profiles of in vitro kinase assays carried out with extracts from wild type and 3 cAMP$^r$ lines. Compared with the wild type activity, the activation of cAPK in CXG-33 and CXG-26 extracts was shifted to a 7 and 220-fold higher cAMP concentration respectively. This class has been called $K_a$ variants and has been shown to reflect mutations in the regulatory subunit of the kinase (Steinberg et al. 1977). The change results in a lowered affinity of the protein for cAMP and the $K_a$ phenotype which has been found to be dominantly expressed (Lemaire and Coffino, 1977). A second class of cAMP$^r$ variants is typified by CXG-22 which exhibits

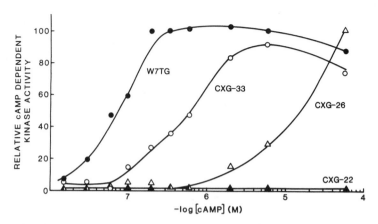

Figure 1. Dose-response of cAMP-dependent protein kinase in extracts of WEHI-7 and 3 cAMP$^r$ cell lines. Cytosol samples were incubated with increasing concentrations of cAMP in the presence of [$^{32}$P]-ATP and histones. The rate of histone phosphorylation was determined as described in Gruol and Dalton (1984). The data have been normalized to the maximum of the wild type activity.

no cAPK activity at all. This phenotype is referred to as kin⁻ and is also dominantly expressed (Steinberg et al., 1978). In the S49 cell line, kin⁻ variants have been shown to lack the kinase catalytic subunit protein (van Daalen Wetters et al., 1983).

We have utilized both the $K_a$ and kin⁻ variants to ask if cAMP regulation plays a role in the lytic response to steroids. We first addressed the capacity to regulate glucocorticoid receptor expression, and found that dexamethasone binding could be significantly (2-fold) increased by exposure of WEHI-7 cells to dibutyryl cAMP (Gruol et al., 1986b). The effect occurred in a dose-dependent fashion, with $K_a$ variants requiring higher concentrations of cyclic nucleotide than wild type cells. The kin⁻ variants did not display any cAMP-dependent change in binding capacity. Figure 2 provides an example of this type of behavior. Both wild type cells and a kin⁻ variant (CXG-56) were treated for increasing times with forskolin (an adenylate cyclase activating drug). At the times indicated, the cells were tested for

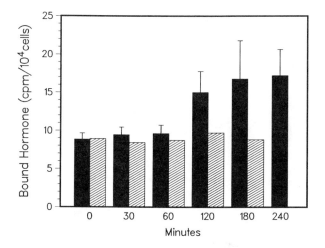

Figure 2. Effect of forskolin on dexamethasone binding to W7TG and CXG-56 cells. Cell cultures were incubated at 37°C with 25 μM forskolin for the indicated times. At the end of the incubation, the cells were collected and suspended in media containing [3H] triamcinolone acetonide (3 X 10⁻⁸M) for 1 hour. The cells were washed 3 times in cold (0°C) phosphate buffered saline and the bound hormone was measured. Parallel cultures containing a 200 fold excess of unlabeled hormone were used to measure non-specific binding. Each point was carried out in duplicate. Solid bars, W7TG (mean and standard deviation of 4 separate experiments); hatched bars CXG-56 (a single experiment).

their capacity to bind dexamethasone. After an initial lag (60 minutes), the steroid binding capacity in the wild type cells nearly doubled by 180 minutes. There was no effect of forskolin on dexamethasone binding in CXG-56 cells.

The results presented above suggest two possibilities. First, activation of cAPK may cause a conversion of preexisting receptors into their active steroid binding form. Alternatively, the observed increase in steroid binding could reflect the synthesis of larger amounts of receptor protein. The 60 minute lag seen prior to observing an increase in binding would appear to support the latter hypothesis, particularly since the subsequent increase in binding is dependent upon continued protein synthesis (Gruol and Bourgeois, 1987).

## WEHI-7

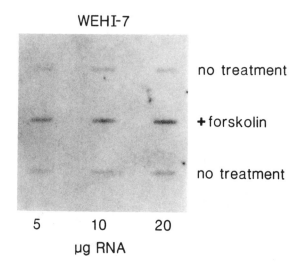

no treatment

+forskolin

no treatment

5      10      20

μg RNA

Figure 3. Slot blot analysis of glucocorticoid receptor mRNA content in WEHI-7. Total RNA was isolated from WEHI-7 cells that had been grown in the presence or absence of 16 $\mu$M for 6 hours. The RNA was bound to nylon filters and hybridized with a [$^{32}$P] cDNA probe to the glucocorticoid receptor gene. Autoradiography was carried out after washing the filters to remove unbound probe.

To address this issue, we have measured the effect of cAMP upon the levels of glucocorticoid receptor mRNA. Figure 3 shows the results of a "slot blot" analysis of RNA from wild type WEHI-7 cells before and after a 6 hour treatment with forskolin. Total cellular RNA was bound to nylon membranes and hybridized with a $^{32}$P-cDNA probe to the human glucocorticoid receptor gene. The autoradiogram shows that forskolin caused an approximate doubling of the mRNA for the receptor. Thus, it seems likely that the increase in steroid binding observed in Figure 2 was the result of increased synthesis of receptor protein.

Lowered Steroid Response in Kinase Defective Variants

Close examination of the cAMP[r] derivatives of WEHI-7 revealed that their response to glucocorticoids was slightly, but measurably, diminished. That is, the variants required higher concentrations of

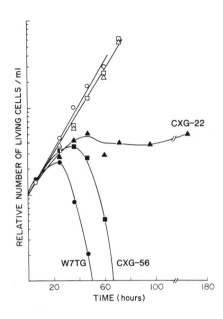

Figure 4. Effect of dexametha-sone on the growth of WEHI-7 and two cAMP-resistant deriva-tives. Cell cultures (5 X $10^4$ cells/ml) were incubated with/without $10^{-6}M$ dexametha-sone for the times indicated. The number of living cells was measured using trypan blue exclusion. The filled symbols represent cells grown in the presence of steroid; the open circle, cells were grown in the absence of steroid.

dexamethasone to initiate a lytic response and a longer interval for complete loss of viability than the wild type cells. This caused us to screen the cAMP[r] cell lines looking for examples which might exhibit a more significant shift in steroid sensitivity. Several, but not all, of the kin⁻ variants were found to be unique in this regard. Examples, illustrating the range of sensitivities that were found, are shown in Figure 4. The data represent the growth (with and without $10^{-6}M$ dexamethasone) of wild type WEHI-7 and 2 kin⁻ variants, CXG-56 and CXG-22. CXG-56 displayed a slightly increased capacity to survive steroid treatment which was observed in most of the cAMP[r] variants. CXG-22, on the other hand, had considerably greater resistance to steroid. In the presence of dexamethasone, the number of CXG-22 cells initially increased 3-fold and then remained essentially constant for the remainder of the experiment (a total of 6 days). This behavior does not represent cell cycle arrest since the number of dead cells was continually increasing in the culture.

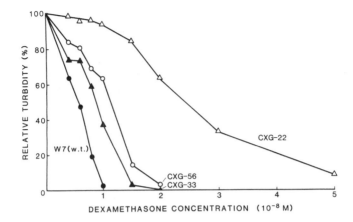

Figure 5. Dose response to dexamethasone of WEHI-7 and 3 cAMP[r] derivatives. Cell cultures (10[3] cells/ml) were incubated with increasing
concentrations of dexamethasone for 10 days. At the end of the incubation period, the turbidities of the cultures were measured (660 nm)
and expressed relative to untreated cultures.

The increased steroid resistance of the CXG-22 cell line is also mani-
fested by the data presented in Figure 5. It shows dose-response
curves reflecting the capacity of cells to proliferate in varying concentrations of dexamethasone. CXG-33 is a $K_a$ variant that was characterized in Figure 1. Both CXG-33 and CXG-56 exhibited an approximately 2-fold decrease in steroid sensitivity, typical of most cAMP[r]
variants. CXG-22 scored as considerably more resistant, nearly 4-fold
higher steroid concentrations were required for killing than was
necessary for wild type cells.

Acquisition of steroid resistance in murine T-cell lymphomas has
heretofore been associated with changes in glucocorticoid receptor
function (steroid binding and nuclear translocation). Table I lists
the results of a series of experiments characterizing cAPK function,
steroid sensitivity and three receptor parameters in a variety of
cAMP[r] cell lines.

Table I.  Phenotypic data from WEHI-7 and cAPK defective variants.

| Cell Line | Receptors Per Cell | $K_D$ ($10^{-8}$ M) | Nuclear Translocation | cAPK Phenotype |
|---|---|---|---|---|
| W7TG (W.T.) | 29,000 | 0.90 | 53.3% | W.T. |
| CX1G-33 (dex$_s$) | 28,000 | 0.85 | 53.7% | $K_a$ - 7X |
| CX1G-41 (dex$^S$) | 27,000 | 0.90 | 51.7% | $K_a$ - 8X |
| CX1G-21 (dex$^S$) | 28,000 | 0.77 | 55.1% | $K_a$ - 66X |
| CX1G-26 (dex$^S$) | 26,000 | 0.57 | 46.0% | $K_a$ - 220X |
| CX1G-22 (dex$^{ir}$) | 27,000 | 1.4 | 51.2% | kin$^-$ |
| CX1G-49 (dex$^{ir}$) | 27,000 | 1.1 | 50.4% | kin$^-$ |
| CX1G-56 (dex$^S$) | 27,000 | 0.89 | 50.0% | kin$^-$ |

The dex$^S$ designation is used for all cell lines that possessed a complete lytic response to dexamethasone ($10^{-6}$M), even if the dose response was slightly shifted.  The designation dex$^{ir}$ is used to signify that the steroid resistance was incomplete.  That is, cell death was observed, but living cells remained in the cultures even after a protracted period of time (>5 days). The results demonstrate that the cAMP$^r$ variants contained receptors with normal capacity to bind hormone and to translocate to the nucleus.  Thus, the increased resistance to steroids observed in the CXG-22 and CXG-49 cell lines does not appear to be due to a change in receptor function.  This suggests that the loss of cAPK  activity may have the potential to alter other components of the autolytic process.


Defective cAPK, an Initial Step towards Steroid Resistance

The fact that cell lines containing defective cAPK activity displayed a muted response to glucocorticoids suggested that loss of kinase function might serve as an initial step towards isolation of a new class of steroid-resistant variants. A large number of glucocorticoid-resistant derivatives of WEHI-7 had been previously isolated and characterized in this laboratory (Pfahl et al., 1978; Pfahl and Bourgeois, 1980; Johnson et al., 1984; Bourgeois and Gasson, 1985).

These studies had found that WEHI-7 is functionally diploid ($r^+/r^+$) for expression of the receptor gene (Bourgeois and Newby, 1977). This is reflected by its 25,000-30,000 dexamethasone binding sites per cell. Partially resistant variants, containing half the normal steroid binding capacity (15,000 per cell), are functionally haploid ($r^+/r^-$) for receptor function. Variants which are totally refractory ($dex^r$) to dexamethasone can be isolated at a high frequency ($10^{-6}$) from the intermediate ($r^+/r^-$) cell lines. This concept can be outlined as follows:

$$r^+/r^+ \longrightarrow r^+/r^- \longrightarrow r^-/r^-.$$

Because of the diploid nature of receptor expression in WEHI-7, isolation of the $dex^r$ phenotype without prior mutagenesis had not proven feasible (Huet-Minkowski et al., 1981). In other words, spontaneous loss of function from both receptor alleles (in a single step) had not been found with a measurable frequency ($<10^{-10}$). These observations place into perspective the fact that we have found that spontaneous $dex^r$ variants can be isolated at a high frequency (approximately $10^{-7}$) from many of the $cAMP^r$ cell lines (Gruol et al., 1986b). Forty nine of the 60 $cAMP^r$ $dex^s$ (Stage I) lines tested so far have yielded $cAMP^r$ $dex^r$ (Stage II) variants when cultures of these cells were incubated with $10^{-6}$M dexamethasone. Thus, defective cAPK activity serves as a permissive background for the generation of dexamethasone-resistant variants at a high frequency. The selection scheme is outlined as follows:

$$\text{Stage I} \qquad\qquad \text{Stage II}$$
$$cAMP^s\ dex^s \longrightarrow cAMP^r\ dex^s \longrightarrow cAMP^r\ dex^r$$
$$\sim 5 \times 10^{-7} \qquad\qquad \sim 10^{-7}$$

Growth curves carried out with the $cAMP^r$ $dex^r$ (Stage II) variants in the presence of dexamethasone showed them to be totally refractory to the hormone (Gruol et al., 1986b). This suggested either a total loss of glucocorticoid receptor function, or a defect at some subsequent step in the lysis pathway. Somewhat surprisingly, characterization of the receptors in the Stage II cell lines revealed a variety of receptor phenotypes. Typical examples are given below.

Table II.  Characterization of Several Stage I and Stage II Variants

| Cell Line | Phenotype | Receptors per Cell | $K_d$ $(10^{-8}M)$ | % Nuclear Receptors | Maximum Receptors per Nucleus |
|-----------|-----------|--------------------|--------------------|---------------------|-------------------------------|
| W7TB | $cAMP^s$ $dex^s$ | 28,000 | 1.4 | 46.6 | 13,000 |
| C1X-46 | $cAMP^r$ $dex^s$ | 21,700 | 1.0 | 42.6 | 9,240 |
| C1X-46D1 | $cAMP^r$ $dex^r$ | 4,400 | 2.37 | 14.7 | 650 |
| C1X-48 | $cAMP^r$ $dex^s$ | 27,700 | 1.3 | 42.4 | 11,700 |
| C1X-48D2 | $cAMP^r$ $dex^r$ | <1,400 | N.D. | N.D. | -- |
| C1X-48D3 | $cAMP^r$ $dex^r$ | 20,500 | 1.2 | 31.4 | 6,440 |
| C1X-48D4 | $cAMP^r$ $dex^r$ | 13,200 | 1.8 | 45.0 | 5,940 |
| C1X-49 | $cAMP^r$ $dex^s$ | 21,700 | 1.5 | 39.6 | 8,590 |
| C1X-49D1 | $cAMP^r$ $dex^r$ | 13,700 | 1.8 | 41.8 | 5,730 |

The steroid binding in the $cAMP^r$ $dex^r$ variants ranged from less than 1,400 to 20,500 sites per cell. C1X-46D1 and C1X-48D2 appear to have lost most of their receptor function in a single step. In contrast, C1X-48D3 had nearly wild type levels of binding and only a partial loss of nuclear translocation (from 47% to 31%).

The far right column of Table II is an estimate of the number of receptors found in the nuclei at saturating concentrations of hormone. The values were calculated from the receptors per cell and the fraction translocated to the nucleus. The cell lines C1X-48D3, C1X-48D4 and C1X-49D1 all had considerble amounts of receptor translocated to the nuclear compartment. We estimate from experiments such as shown in Figure 5, that 5 to 6 thousand receptors per nucleus are sufficient to produce an effective cytolytic response, even in the $K_a$, $cAMP^r$ lines. Thus, the failure of the Stage II variants (C1G-48D3, C1X48D4 and C1X49D1) to initiate an autolytic response indicates a loss of function at a level that resides beyond the ability of receptor to bind hormone and translocate to the nucleus.

Glucocorticoid-Induced Responses in WEHI-7 Cell

The cytolytic process triggered by glucocorticoids in WEHI-7 represents a model for steroid-regulation of specific genes in thymocytes. Current efforts in this field are directed towards identifying the changes that are induced and understanding their role in triggering the cellular "suicide" mechanism. Recent work has focused attention upon the effects of glucocorticoids on the integrity of the nucleus and its chromatin components. Steroid mediated increases in T-cell nuclear fragility (Nicholson and Young, 1978), chromatin condensation (Burton et al., 1967) and DNA fragmentation (Wyllie, 1980; Duke et al., 1983) have all been shown to be early events (within 6 hours) in the cellular response. This time frame is within the onset of lysis. These observations have led to the hypothesis that glucocorticoids either induce the synthesis of a lethal nuclease or cause the activation of preexisting nucleases. Both possibilities have been tested experimentally.

Work carried out in this laboratory has confirmed that glucocorticoids promote DNA fragmentation in T-cell lymphomas. The results have also produced evidence for preexisting nucleases within the nuclei of WEHI-7 cells. This is consistent with the observations made by Cohen and Duke (1984) who suggested that glucocorticoid treatment caused activation of existent nucleases which led to the DNA degradation observed during lysis. Our experience gained from attempts to verify the alterative hypothesis, induced de novo synthesis of a nuclear nuclease, has caused us to withhold judgement of some of the published data supporting this possibility. Experiments carried out by Compton and Cidlowski (1987) employed salt extraction of isolated nuclei and subsequent analysis for solubilized nuclease activity. They found an apparent induction (increased concentration in solution) of a family (3-4 species) of low molecular weight nucleases released from the nuclei of steroid treated cells. This approach has a number of potential difficulties, however, since nuclei from steroid treated cells contain fragmented DNA. This condition promotes more efficient release of chromatin proteins during salt extractions. A second problem lies in the method that was used to detect nuclease activity. The protocol, as described by Rosenthal and Lacks (1977), relies on SDS-PAGE gels impregnated with DNA and stained with ethidium bromide. Nucleases separated in this way are then detected by their ability to digest the included DNA. We found that DNA-binding proteins such as histones could also score positively in this system by virtue of their capacity to exclude the ethidium bromide stain. Therefore, we believe that the induced synthesis of nucleases by glucocorticoids in T-cells is still very much an open question. Its resolution will require a refinement of the current methods or the use of alternative approaches.

This laboratory has opted to pursue a different and broader strategy in addressing the nature of programmed cell death in T-lymphocytes. We have chosen to employ a molecular approach by cloning those genes which are selectively expressed in response to a combination of steroids and cAMP. Under these conditions, specific subsets of genes will be induced by each drug alone or by their combination. Thus, the potential exists to identify genes which participate not only in each arm of the lytic pathway, but those gene products which also reflect important points of intersection between the two signaling systems.

Recent advances in molecular cloning technology have provided a means of isolating genes which are differentially expressed. In this instance, RNA prepared from treated (dexamethasone, cAMP) wild type WEHI-7 cells and untreated Stage II variants (cAMP$^r$, dex$^r$) differ by their induced gene content. Through a process of subtractive hybridization, a cDNA probe enriched for induced gene sequences can be generated and used to screen a cDNA library made from RNA of treated wild type cells (Davis et al., 1984). Our initial efforts in this regard have been encouragingly fruitful, and we are now in the process of characterizing a number of different cAMP/steroid responsive clones. To our knowledge, these represent unique examples of steroid and cyclic nucleotide regulated genes in T-cells.

DISCUSSION

There is considerable evidence to indicate that glucocorticoid hormones affect cAMP metabolism in a variety of cell types (Davies and Lefkowitz, 1984). Steroid effects on $\beta$-adrenergic receptor levels (Foster and Harden, 1980; Fraser and Venter, 1980; Collins et al., 1988), phosphodiesterase activity (Manganiello and Vaughan, 1972), and more recently, the levels of G-protein ($\alpha$-subunit) expression (Chang and Bourne, 1987) have all been reported. Steroids may also play a role in regulating the covalent modification of the cAMP-dependent protein kinase regulatory subunits (Liu, 1984). There has been far less to indicate that cAMP regulation might feed back upon the glucocorticoid receptor signaling system. There have been several reports indicating that steroid receptors are substrates for cAPK in vitro, (Singh and Moudgil, 1985; Weigel et al., 1981) but there is no evidence that this process takes place in intact cells.

Murine T-cell lymphomas cell lines such as WEHI-7 and S49 offer a unique opportunity to ask if cAMP regulation interfaces with steroid receptor-mediated control. Since both signaling systems provoke a cytolytic response, resistant variants containing defects in each pathway can be obtained. Our strategy has been to try to combine what had been two separate lines of study. We have asked if changes in cAPK function alter steroid responsiveness. This question had been addressed previously by Gehring and Coffino (1977) who found no change of steroid sensitivity in cAMP$^r$ variants of S49. However, these authors had looked for altered response to high ($10^{-7}$M) concentrations of dexamethasone. We have assayed for more subtle differences which might reflect changes in a multifaceted response to steroids that overlap with the cAMP-mediated pathway. In this regard, it is valuable to point out that the two cytolytic pathways are, in all likelihood, not dependent upon one another. SAK-8, a lysis defective mouse cell line containing normal glucocorticoid receptors (Gasson and Bourgeois, 1983a), is still sensitive to cyclic nucleotide-mediated killing (unpublished results). This does not rule out the possibility, however, that the two pathways share common elements. The data shown in Figures 4 and 5 clearly demonstrate that the cAPK status can have a profound effect on steroid-mediated processes in T-cells.

Loss of cAPK activity did not result in a measurable loss of glucocorticoid receptor function (Tables I and II). Even though cAMP can elicit an increase in steroid binding, the kin⁻ cAMPʳ variants continued to contain steroid binding levels near or at wild type capacity. This suggests that the role of cAMP regulation in glucocorticoid receptor expression may only be realized during brief periods of elevated intracellular cAMP levels. This could be part of a cell cycle rhythm, higher levels of glucocorticoid binding have been found during S phase in HeLa cells (Cidlowski and Cidlowski, 1982). Alternatively, the additional cAMP-stimulated binding could represent part of the programmed response to stress. Catecholamines, through their interaction with the $\beta$-adrenergic receptor system, may temporarily promote greater glucocorticoid receptor expression, thus insuring a heightened response to the adrenal steroids.

Early experiments with thymocytes demonstrated a requirement for gene activation and protein synthesis to sustain a steroid-mediated lytic response. An extensive analysis of gene induction using 2-D SDS-PAGE has subsequently revealed that a limited set (6 out of 2500) of the T-cell genes were capable of responding to hormone treatment (Voris and Young, 1981). These results testify to the specificity of the changes and suggest that the genes responsible for cytolysis may be few in number. This situation should, therefore, be amenable to a molecular cloning approach. We are proceeding in that direction, but have broadened the search to include the possible involvement of the cAMP signaling pathway. The first step is well underway, isolating genes that are dexamethasone/cAMP responsive. These are also likely to include genes that play an important role in lymphoid cells, but are not involved in the lytic pathway. Identifying all of the gene products and establishing which ones are involved in cytolysis will be more challenging.

## Acknowledgements

We thank Matthew N. Ashby, N. Faith Campbell, Dyana K. Dalton and Fatemah M. Rajah for their excellent technical assistance. We also acknowledge the valued contribution of Gloria Laky Swart in the transcription of the manuscript. The work was carried out with the support of NIH grants CA36146 (S.B.), DK38131 (D.G.), and of a grant from the Elsa U. Pardee Foundation (S.B.). M.T.H. was supported by NIH Training Grant CA09254 to Dr. Melvin Cohn.

## References

Baxter, J.D., and Rousseau, G.G. (1979) In: Monographs on Endocrinology: Glucocorticoid Hormone Action, (Baxter JD, Rousseau GG, eds) Springer-Verlag, Berlin 12, pp 1-24.

Bourgeois, S., and Gasson, J.C. (1985) Biochemical Actions of Hormones, Vol XII, pp 311-351.

Bourgeois, S., and Newby, R.F. (1977) Cell 11, 423-430.

Burton, A.F., Storr, J.M., and Dunn, W.L. (1967) Canadian. J. Biochem. 45, 289-297.

Chang, F.-H., and Bourne, H.R. (1987) Endocrinology 12, 1711-1715.

Cidlowski, J.A., and Cidlowski, N.B. (1982) Endocrinology 110, 1653-1662.

Claman, H.N. (1972) New Eng. J. Med. 287, 388-397.

Coffino, P., Bourne, H.R., Friedrich, U., Hochman, J., Insel, P.A., Lemaire, I., Melmon, K.L., and Tomkins, G.M. (1976) Recent Prog. Horm. Res. 32, 669-684.

Cohen, J.J., and Duke, R.C. (1984) J. Immunol. 132, 38-42.

Collins, S., Caron, M.G., and Lefkowitz, R.J. (1988) J. Biol. Chem. 263, 9067-9070.

Compton, M.M., and Cidlowski, J.A. (1987) J. Biol. Chem. 262, 8288-8292.

Davies, A.O., and Lefkowitz, R.J.(1984) Ann. Rev. Physiol. 46, 119-130.

Davis, M.M., Cohen, D.I., Nielsen, E.A., Steinmetz, M., Paul, W.E., and Hood, L. (1984) Proc. Natl. Acad. Sci. USA 81, 2194-2198.

Duke, R.C., Chervenak, R., and Cohen, J.J. (1983) Proc. Natl. Acad. Sci. USA 80, 6361-6365.

Foster, S.J., and Harden, T.K. (1980) Biochem. Pharmacol. 29, 2151-2153.

Fraser, C.M., and Venter, J.C. (1980) Biochem. Biophys. Res. Commun. 94, 290-397.

Gasson, J.C., and Bourgeois, S. (1983a) J. Cell. Biol. 96, 490-415.

Gasson, J.C., and Bourgeois, S. (1983b) In: UCLA Symposia on Molecular and Cellular Biology: Rational Basis for Chemotherapy, (Chabner BA, ed) Alan R Liss, Inc, New York, pp 153-176.

Gasson, J.C., Ryden, T., and Bourgeois, S. (1983) Nature 302, 621-623.

Gehring, U., and Coffino, P. (1977) Nature 268, 167-169.

Granner, D.K., and Hargrove, J.L. (1983) Mol. Cell. Biochem. 53-54, 113-128.

Gruol, D.J., Ashby, M.N., Campbell, N.F., and Bourgeois, S. (1986a) J. Steroid Biochem. 24, 255-258.

Gruol, D.J., and Bourgeois, S. (1987) In: Recent Advances in Steroid Hormone Action (Moudgil VK, ed) Walter de Gruyter & Co, New York, pp 315-335.

Gruol, D.J., Campbell, N.F., and Bourgeois, S. (1986b) J. Biol. Chem. 261, 4909-4914.

Gruol, D.J., and Dalton, D.K. (1984) J. Cell. Physiol. 119, 107-118.

Harmon, J.M., Norman, M.R., Fowlkes, B.J., and Thompson, E.B. (1979) J. Cell. Physiol. 98, 267-278.

Housley, P.R., and Pratt, W.B. (1983) J. Biol. Chem. 258, 4630-4635.

Huet-Minkowski, M., Gasson, J.C., and Bourgeois, S. (1981) Cancer Res. 41, 4540-4546.

Johnson, D.M., Newby, R.F., and Bourgeois, S. (1984) Cancer Res. 44, 2435-2440.

Kalinyak, J.E., Dorin, R.I., Hoffman, A.R., and Perlman, A.J. (1987) J. Biol. Chem. 262, 10441-10444.

Lemaire, I., and Coffino, P. (1977) J. Cell. Physiol. 92, 437-446.

Liu, A.Y.-C. (1984) Trends Biochem. Sciences 5, 106-108.

Manganiello, V., and Vaughan, M. (1972) J. Clin. Invest. 51, 2763-2767.

Mendel, D.B., Bodwell, J.E., and Munck, A. (1987) J. Biol. Chem. 262, 5644-5648.

Munck, A., and Crabtree, G.R .(1981) In: Cell Death in Biology and Pathology (Bowen ID, Looksin RA, eds) Chapman and Hall Ltd., New York pp 329-353.

Nicholson, M.L., and Young, D.A. (1978) Cancer Res. 38, 3673-3680.

Okret, S., Poellinger, L., Dong, Y., and Gustafsson, J.-A. (1986) Proc. Natl. Acad. Sci. USA 83, 5899-5903.

Pfahl, M., and Bourgeois, S. (1980) Somatic Cell Genet. 6, 63-74.

Pfahl, M., Kelleher, Jr. R.J., and Bourgeois, S. (1978) Mol. Cell. Endocr. 10, 193-207.

Rajah, F.M., Gruol, D.J., and Bourgeois, S. (1988) FASEB J. 2, A590.

Rosenthal, A.L., and Lacks, S.A. (1977) Anal. Biochem. 80, 76-90.

Rosewicz, S., McDonald, A.R., Maddux, B.A., Goldfine, I.D., Miesfeld R.L., and Logsdon, C.D. (1988) J. Biol. Chem. 263, 2581-2584.

Short, J.M., Wynshaw-Boris, A., Short, H.P., and Hanson, R.W. (1986) J. Biol. Chem. 261, 9721-9726.

Singh, V.B., and Moudgil, V.K. (1985) J. Biol. Chem. 260, 3684-3690.

Steinberg, R.A., O'Farrell, P.H., Frederick, U., and Coffino, P. (1977) Cell 10, 381-391.

Steinberg, R.A., van Daalen Wetters, T., and Coffino, P. (1978) Cell 15, 1351-1361.

Thompson, E.B., and Lippman, M.E. (1974) Metabolism 2, 159-202.

van Daalen Wetters, T., Murtaugh, M.P., and Coffino, P. (1983) Cell 35, 311-320.

Voris, B.P., and Young, D.A., (1981) J. Biol. Chem. 256, 11319-11329.

Weigel, H.L., Tash, J.S., Means, A.R., Schrader, W.T., and O'Malley, B.W. (1981) Biochem. Biophys. Res. Com. 102, 513-519.

Wyllie, A.H., (1980) Nature 284, 555-556.

Young, D.A., Nicholson, M.L., Voris, B.P., and Lyons, R.T. (1980) In: Hormones and Cancer, (Iacobelli S, et al. ed) Raven Press, New York, pp 135-155.

DISCUSSION OF THE PAPER PRESENTED BY S. BOURGEOIS

MOUDGIL: Assuming some of the cAMP effects on glucocorticoid receptor or gene expression are direct, have you looked at phosphorylation in vivo of glucocorticoid receptor directly?

BOURGEOIS: We have not yet assessed the effect of cAMP upon the level of receptor phosphorylation in WEHI-7.

SIMONS: Have you looked at the effects of cAMP in the presence of phosphodiesterose inhibitors?

BOURGEOIS: Methylisobutylxanthine (MIX), when added with forskolin, caused a slight enhancement of the increase in steroid binding. In fact, when added alone, MIX was also capable of producing a significant increase in glucocorticoid binding capacity.

SIMONS: Have you been able to look at any of the physical properties of the receptors in Dex$^R$ cells by Western blots, or of the receptor mRNAs by Northern blots?

BOURGEOIS: We are currently engaged in analyzing a series of the variants using the Northern blot technique.

SIMONS: It appeared that the expression of pT1.5 is greater in the double mutant (Dex$^R$ cAMP$^R$) than in the wild type cells. Would you care to comment on this. Also, how were you able to isolate this cDNA in your subtractive hybridization selection if this mRNA is present in your "uninduced" cells?

BOURGEOIS: There is a basal level of expression of pT1.5 mRNA in the double mutant but we have not quantitated this mRNA, and are not convinced that it is higher than the basal level in wild type cells. We were able to isolate this cDNA by subtractive hybridization because the induced level of this mRNA in wild type cells is greatly increased.

CIDLOWSKI: Is your turbidity assay actually not a measure of inhibition of cell growth rather than a stimulation of cell death?

BOURGEOIS: It is a measure of both since we have not found that WEHI-7 cells remain viable when they become cell cycle arrested. In addition, visual inspection of the cultures has consistently confirmed the loss of viability.

TATA: Have you carried out your Northern analyses on RNA extracted as a function of time after treatment with Dexamethasome and cylic AMP?

BOURGEOIS: Those experiments are in progress.

TATA: Do you think that the mRNAs picked up by your cloned probes may be secondarily induced by a small level change in a primary product?

BOURGEOIS: This is certainly a possibility since the cells used to construct the cDNA library were treated with drugs for 5 hours. We are testing the effect of inhibitors of protein synthesis on mRNA induction to distinguish primary from secondary responses.

TATA: Did you carry out nuclear run-off assays with the cloned probes?

BOURGEOIS: No.

ROY: Did you examine the effect of non-glucocorticoid/non-cAMP lytic agents on the induction of these mRNAs?

BOURGEOIS: I am not aware of lymphocytolytic agents other than glucocorticoids and cAMP. We have not tested the effect of other types of toxic drugs on the induction of these mRNAs

DISCUSSANTS: V. MOUDGIL, S. BOURGEOIS, S.S. SIMONS, J. CIDLOWSKI, J.R. TATA AND A.K. ROY

# DIFFERENTIAL REGULATION OF TYROSINE AMINO-TRANSFERASE BY GLUCOCORTICOIDS:

## TRANSCRIPTIONAL AND POST-TRANSCRIPTIONAL CONTROL

E. Brad Thompson , P. Gadson , G. Wasner  and S.S. Simons, Jr.

Current models of steroid hormone action focus on three components:  the ligands (steroids) themselves, their receptors, and the cis-active steroid response elements (SRE's) to which the receptor binds to cause increased or decreased transcriptional activity from nearby genes. Many studies have found that as steroid concentration increases across the range that leads to full occupancy of the receptor, a constant fraction of the steroid-receptor complex becomes more tightly associated with nuclei, presumably interacting with certain SRE's to effect increasing degrees of response (Baxter et al., 1979). The foundations of these concepts were first established with observations on the induction of tyrosine aminotransferase (E C 2.6.1.5) in HTC cells. This enzyme was one of two that in the late 1950's had been observed to increase in rat liver in response to glucocorticoid administration (Lin et al., 1957; Lin et al., 1958; Greengard et al.). By establishing a line of liver-derived hepatoma cells (HTC cells), it was possible to resolve clearly several issues not easy to work out in whole animal or whole-organ systems (Thompson et al., 1966). Among these were the fact that in HTC cells, the degree of induction of the transaminase corresponded to the degree of occupancy of the receptor by active glucocorticoids (Baxter et al., 1970; Baxter et al., 1971; Rousseau et al., 1972). This principle was found to hold for a large number of inducible genes, driven by various types of steroids and their receptors.

The value of any model in science lies in its ability to explain existing data and to predict responses. The current model of steroid hormone action is important because of its ability to fit with a wide body of existing data. There is no doubt that it does so, and it is within the context of that model that much current research is done. The danger of a powerful model, however, is that it can restrict thinking by encouraging workers to exclude or avoid examination of data that fails to fit. It is always important, therefore, to carefully establish data that do not fit any existing model and to verify and explore them. If the model cannot explain them, it must then be modified or replaced.  Consequently, we have returned to the tyrosine aminotransferase system to examine how well the general model explains certain observations.  In recent years considerable new information has been obtained about glucocorticoid induction of tyrosine aminotransferase. Schütz and his collaborators have cloned its hepatic gene, have shown that induction by glucocorticoids increases the quantity of its messenger RNA, and have provided data consistent with

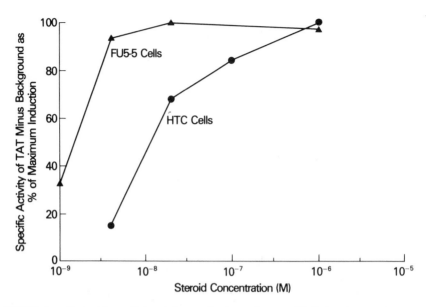

FIGURE 1: Induction of tyrosine aminotransferase (TAT) by dexamethasone in HTC and Fu5-5 cells. The increase in TAT specific activity above the basal level is plotted as per cent of the maximum increase seen for HTC (●) or Fu5-5 (▲) cells [SA of TAT for HTC, 17.2 (basal) and 124 (induced); for Fu5-5, 83.3 (basal) and 252 (induced)]. Each point represents the average of duplicate dishes of cells. Reprinted with permission from (Mercier et al.).

the interpretation that at least part of this increase is due to increased transcription from the gene (Scherer et al., 1982; Schmid et al., 1987). They have sequenced the promotor region of tyrosine aminotransferase and have found several glucocorticoid response elements far upstream of the transcription start site. Their many ingenious experiments utilizing in vitro mutants and molecular cloning techniques have shown the importance of these sites for the induction of the gene, despite the fact that they are more than two kilobases upstream from the actual start of transcription (Jantzen et al.). Indeed, one of the enigmas of this system is the lack of a ready explanation of how these hormone responsive enhancer-like elements can act so very far from the actual transcription initiation site.

Tyrosine aminotransferase induction in Fu5-5 cells does not follow the predicted response to induction by glucocorticoids

The newandunexpectedobservationthatdemandedattention concerned the dose-response curves for the induction of tyrosine aminotransferase by glucocorticoids in a pair of rat hepatoma cell lines, HTC and Fu5-5. The origin of these two lines shows that they are clearly both liver derived but from different sources. HTC cells were originally grown from the ascites form of Morris Hepatoma 7288c raised in the inbred Buffalo rat (Thompson et al., 1966). Fu5-5 cells were cloned by Weiss and collaborators from the Rueber H35 hepatoma cell line, originally induced in the AxC male rat, not an inbred strain (Weiss et al.). Both hepatomas had been induced in the animals by chemical carcinogens. The basic dose-response data in the two cells is shown in figure 1. As the concentration of steroid is raised, one sees that the induction of tyrosine aminotransferase in the Fu5-5 cell occurs at considerably lower concentrations of hormone than that in the HTC cells. There is a six to ten fold greater sensitivity for induction in the Fu5-5 cells. Over a period of eight years, Simons and collaborators found that although the basal level of tyrosine aminotransferase in both HTC and Fu5-5 cells varies somewhat with time, the dose-response interval for induction of the aminotransferase remains constant at about six-fold between them. That is Fu5-5 cells, regardless of the basal level of enzyme activity, always are more sensitive to glucocorticoids for the induction of tyrosine aminotransferase (Simons et al., 1988a,b). By the standard model, focussing on the glucocorticoid receptors, one might expect to find that the receptor in Fu5-5 cells was either more plentiful or had a higher affinity for the ligand. This however is not the case. The number of receptors per cell in Fu5-5 and HTC cells is approximately the same, with more receptors if anything in HTC cells than in Fu5-5. The receptors are also quite similar in their affinity for glucocorticoids (table 1). Other standard measurements of receptor physical properties in the two cell lines (i.e., $M_r$, isoelectric point, protease fragmentation pattern, and strength of binding to DNA) show them to be quite similar (Miller et al.). Another possibility that might account for the peculiar sensitivity of Fu5-5 cells would be that for some reason the receptor-ligand complex in them associates more readily with the nucleus and the DNA therein. In other words, there might be more efficient "nuclear translocation" at a given concentration of hormone in Fu5-5 cells, allowing a greater concentration of activated hormone receptor complexes to exist in the nucleus, acting to enhance transcription of the aminotransferase gene. This, however, also is not the case. Both cells show approximately the same degree of nucleus-associated steroid receptor complexes at each of the critical inducing concentrations of hormone (table 2). Thus the simple quantity of association of the steroid receptor complex with nucleus does not differ significantly in the two cell lines and cannot account for the greater sensitivity of Fu5-5 cell tyrosine aminotransferase induction. One can see at once that the routine receptor-oriented model would fail to predict the greater sensitivity of tyrosine aminotransferase induction in Fu5-5 cells.

It could be that the Fu5-5 cell is generally more sensitive to glucocorticoids than the HTC cell. The sensitivity to glucocorticoids of the cis-acting glucocorticoid response element of the mouse mammary tumor virus, for example, has been shown to vary between certain cell lines (Pfahl et al.). If Fu5-5 cells were generally more sensitive to glucocorticoids, then all glucocorticoid inducible genes in them should show the same leftward shift in the dose response curve as tyrosine aminotransferase. Two genes have been examined carefully in this respect: glutamine synthetase and mouse mammary tumor virus. We found that glutamine synthetase induction in both Fu5-5 and HTC cells follows the same traditional dose-response curve as does the aminotransferase in the HTC cell (figure 2). That is, the extent of response in induction of glutamine synthetase is close to the degree of occupancy of receptor by hormone, and at the Kd of the steroid receptor complex - that concentration at which about 50% of the receptors are occupied by hormone - about 50% of the maximum level of glutamine synthetase induction was reached. But tyrosine aminotransferase induction behaved that way only in the HTC cell; in the Fu5-5 cell, its induction was much greater than 50% at the concentration of steroid capable of saturating half the receptors (figures 1 & 2). Thus it seems clear that for some reason the tyrosine aminotransferase gene is specifically more sensitive in Fu5-5 cells. This was further confirmed by use of mouse mammary

TABLE 1: Affinity of dexamethasone for HTC and Fu5-5 cell receptors

| Type of assay | $K_d$ of steroid-receptor binding | | | | | |
| --- | --- | --- | --- | --- | --- | --- |
| | In HTC cells | | In Fu5-5 cells | | $K_d$ ratio HTC/Fu5-5 | |
| Cell-free at 0 C[a] | 3 h | 20 h | 3 h | 20 h | | |
| | $2.7 \times 10^{-9}$ (2) | $1.1 \times 10^{-9}$ | $3.0 \times 10^{-9}$ (3) | $8.4 \times 10^{-10}$ | 0.91 | 1.31 |
| Whole cell at 37 C[b] | 30 min | 20 h | 30 min | 17–20 h | | |
| With WO/5-5 | $2.8 \times 10^{-8}$ (2) | $1.6 \times 10^{-8}$ (1) | $1.1 \times 10^{-8}$ (3) | $7.5 \times 10^{-9}$ (2) | 2.64 | |
| With WO/5-5 | | | | $6.4 \times 10^{-9}$ (3) | 2.14 | |
| With WO/10 | | | | | | |

The cell free affinities were determined as previously described (Pfahl et al.). The specifically bound [3H] dexamethasone was calculated as total minus nonspecific binding. The data were then plotted by the method of Scatchard and the best straight line was determined by linear regression analysis. The number of experiments is shown in parentheses. $K_d$, Equilibrium dissociation constant.

[a]Average receptor concentration, 0.51 pmol/mg protein for HTC cells and 0.74 pmol/mg protein for FU5-5 cells.

[b]Average receptor concentration, 0.27 pmol/mg protein for HTC cells and 0.41 pmol/mg protein for FU5-5 cells. WO5-5 and WO/10 refer to media used for culture. See original article for details.

[c]Reprinted with permission from (Mercier et al.).

[d]Subsequent whole-cell assays of GR sites showed 84000 sites/cell in HTC and 60000 in FU5-5. Cell volumes of the two lines are similar.

Table 2:  Effect of dexamethasone on TAT gene induction

| (DEX)nM | BINDING[a] SITES (x 10$^{-3}$) | % NUCLEAR[a] TRANSLOCATION | TAT TRANSCRIPTION PPM | RELATIVE[b] TAT mRNA INDUCTION |
|---|---|---|---|---|
| **Fu5-5** | | | | |
| 0 | -- | -- | 400 + 105 | 3 |
| 10 | 32 | 87 | 1020 + 240 | 18 + 2 |
| 100 | 60 | 86 | 1260 + 106 | 22 + 1 |
| **HTC** | | | | |
| 0 | -- | -- | 210 + 122 | 1 |
| 10 | 38 | 87 | 350 + 140 | 4 + 1 |
| 100 | 84 | 78 | 1350 + 220 | 10 + 2 |

**a** Whole cells were treated with various concentrations of [$^3$H] dexamethasone (DEX) in the presence and absence of unlabeled dexamethasone for one hour at 37°C, washed with phosphate-buffered saline, and homogenized. Nuclei were purified from the cell homogenate. Binding sites associated with purified nuclei represents amount of radioactivity bound in the presence and absence of unlabeled competitor ligand. Nuclear translocation is expressed as the percentage of total cell binding sites bound to nuclei.

**b** The level of the uninduced TAT mRNA was arbitrarily set at values corresponding to the level of TAT mRNA in Fu5-5.

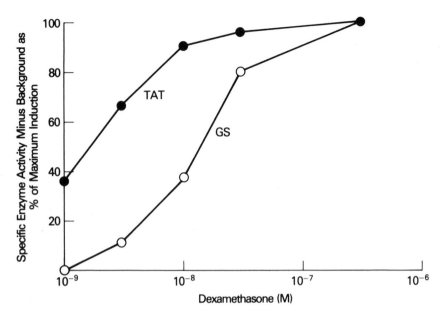

FIGURE 2: Induction of tyrosine aminotransferase (TAT) and glutamine synthetase (GS) by dexamethasone in Fu5-5 cells. The increase in TAT (●) or GS (○) specific activity above the basal level was plotted as per cent of the maximum observed increase. Each point represents the average of duplicate dishes of cells. Reprinted with permission from (Mercier et al.).

tumor virus. As is well known, this virus contains in its long terminal repeat a set of glucocorticoid response elements; in fact it was in this gene that they were originally discovered (Lee et al.). The induction of mouse mammary tumor virus has been shown to be driven by these elements, stimulated by glucocorticoid receptor complexes, and the elements have been used to confer glucocorticoid inducibility to a variety of other genes (Lee et al.; Ringold). We therefore chose MMTV as the classic example of a transcriptionally regulated, glucocorticoid response element-driven system. Fu5-5 cells were infected with mouse mammary tumor virus, and clones with a low number of stably incorporated viruses chosen for examination (Wasner et al., 1988). When a dose-response curve was carried out on these cells, it was found that induction of the virus followed the classic model, just as did glutamine synthetase induction in both Fu5-5 and HTC cells and as tyrosine aminotransferase induction in HTC cells: receptor occupancy corresponded to the degree of induction in a one to one fashion (figure 3). Again it seemed that something about the tyrosine aminotransferase gene only in Fu5-5 cells resulted in extraordinary sensitivity to glucocorticoids.

We next sought to see at what level this sensitivity occurred. Measurements of tyrosine aminotransferase messenger RNA by Northern blotting show that increased mRNA is produced with the same dose-response curve as enzyme activity. This is highlighted by the data of table 3, which shows that at 10uM dexamethasone TAT mRNA is nearly fully induced in Fu5-5 cells whereas in HTC cells only 40% of maximal

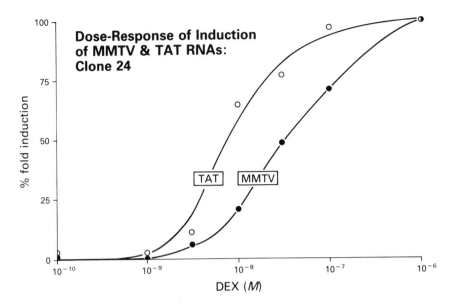

FIGURE 3: Dose-response of induction of TAT and MMTV mRNAs by glucocorticoid. Cells were induced with different concentrations of dexamethasone for 6 hours and harvested for preparation of cytoplasmic extracts. Duplicate blots were hybridized with either TAT-cDNA (o) or MMTV-DNA (●) and subjected to radioautography. The percent maximal induction was calculated for each point from duplicate cultures. A representative experiment for clone 24 is shown. Reprinted, with permission, from (Wasner et al., 1988).

induction has been reached. Therefore the accumulation of messenger RNA is the level at which the differential sensitivity occurs. To pursue this point further, nuclear "run-off" assays were carried out; these provide a measure of the transcriptional activity stimulated in a cell by an inducing ligand. HTC and Fu5-5 cells grown in parallel were exposed to various concentrations of dexamethasone, nuclei prepared, and "run-off" assays of tyrosine aminotransferase transcription carried out (table 4). We found that the Fu5-5 cells exhibited roughly a log more sensitivity to dexamethasone induction of transcription than did HTC cells. Thus at $10_{-8}$M dexamethasone Fu5-5 cells already showed a considerable increase in transcription of the tyrosine aminotransferase gene (after 4 hr exposure), whereas HTC cells showed little change from basal levels. As concentrations of hormone increased, the percent maximal induction of HTC cells caught up with the Fu5-5 cells, and by a high concentration of inducer such as $10_{-6}$M, both were fully activated transcriptionally. We conclude that the tyrosine aminotransferase gene in the Fu5-5 cells is more sensitive to glucocorticoids, at least in part due to greater sensitivity at the transcriptional level. Fu5-5 cells are not only more sensitive to dexamethasone; they also are more sensitive to a variety of other glucocorticoids, and indeed in them anti-hormone analogs which act as anti-glucocorticoids in other cells are partial agonists (Simons et al., 1988b). In addition, Fu5-5 cells show greater sensitivity for induction of tyrosine aminotransferase by cyclic AMP analogs than do HTC cells (Wasner & Simons).

Table 3: Relative tyrosine aminotransferase mRNA levels in Fu5-5 and HTC cells after dexamethasone stimulation.

| Cell | Dex(nM) | % Maximal mRNA TAT |
|------|---------|---------------------|
| Fu5-5 | 0 | 14 |
| | 10 | 82 ± 9 |
| | 100 | 100 |
| HTC | 0 | 10 |
| | 10 | 40 ± 10 |
| | 100 | 100 |

## The extraordinary sensitivity of the tyrosine aminotransferase gene in Fu5-5 cells implies interaction of components

Formally, a leftward shift in sensitivity, such as that we have documented in Fu5-5 cells for the tyrosine aminotransferase gene, can be explained by a system involving a second messenger. It is commonly seen, for example in peptide hormone driven genes, since the peptides act at cell surface receptors and require second message systems to Parallel cultures of Fu5-5 and HTC cells were treated with the indicated concentrations of dexamethasone (Dex). RNA extracted from the cells was quantified for tyrosine aminotransferase messenger RNA (mRNA TAT) by methods previously described (Strobl et al.) using the TAT EH.95 probe kindly provided by G. Schütz, (Scherer et al.; Schmid et al.). Data from (Gadson et al.).

effect their responses.    In such systems, "spare receptors" are often seen, that is, maximal response occurs before all receptor sites are occupied by ligand. Strickland and Loeb have presented mathematical analyses showing that except under extraordinary circumstances, such second message systems mandate that there will be less than a one to one correspondence between receptor occupancy and final response (Loeb and Strickland; Strickland and Loeb).  For most genes in most cells, it seems at present that the steroid hormones do not require an explanation in terms of such a second messenger system, at least not one directly analogous to that of the peptide hormones.  However, the formal argument can be made that for the tyrosine aminotransferase gene in the Fu5-5 cells, some other element in addition to the receptor is involved in the induction.  This could even be at the level of a direct action of the receptor with the promotor region of the gene, if interaction with the component(s) of another control factor acting on that promotor influenced the induction by the steroid.  Direct sequencing and exploration of the Fu5-5 cell tyrosine aminotransferase promotor will be required to explore the possibilities in detail.  Since it is now clear that promotor regions of genes are constellations of regulatory cis-active elements and that the interaction of these elements and the regulatory proteins that bind to them  plays an important role in the

Table 4: TAT gene transcription rate with time after dexamethasone in
HTC and Fu5-5 cells

TAT transcription (ppm)

| Time | $10^{-8}$M Dex | | $10^{-6}$M Dex | |
|------|------|--------|------|--------|
| (hr) | HTC | Fu5.5 | HTC | Fu5.5 |
| 0 | 1.1 | 2.1 | --- | ---- |
| 0.5 | 1.9 | 2.7 | 3.8 | 4.3 |
| 1.0 | 2.4 | 4.6 | 14.6 | 6.2 |
| 2.0 | 3.8 | 5.1 | 6.5 | 10.8 |
| 4.0 | 3.0 | 11.1 | 6.5 | 11.6 |
| 8.0 | --- | ---- | 2.7 | 3.8 |

expression of any particular gene, it may well be that the tyrosine
aminotransferase promotor in Fu5-5 cells contains another regulatory
element which interacts with the its glucocorticoid response element
sites, resulting in the high sensitivity of induction of these cells.
It is also possible to imagine a promotor with tighter specific binding
of receptor-steroid complexes to GRE sites, although current structure-
TAT Gene Transcription Rate with Time in HTC and Fu5-5 Cells. _In vitro_
transcription measurements of HTC and Fu5-5 cells. Cells were treated
with 10 nM or 1000 nM dexamethasone for 0.5, 1, 2, 4, and 8 hours. The
nuclei were isolated and nuclear RNA was labeled _in vitro._ Labeled
nuclear RNA was extracted, transferred to nitrocellulose filters
containing 2 μg each of bound TAT-specific DNA or, as controls, actin
cDNA or PBR322 plasmid. The efficiency of hybridization was
approximately 0.40. The dot blots were analyzed by scintillation
spectrophotometry, and the average results from eight experiments are
shown. From (Gadson et al.).

function analyses do not provide the basis for predicting exactly what
DNA sequences would do so. Furthermore, the cellular milieu in which a
given regulatory element exists can strongly influence the response of
that cell to that regulatory element. Studies on enhancers in general
have shown that they are cell specific to some extent (Schöler and
Gruss; Fischer and Maniatis) and specifically the mouse mammary tumor
virus glucocorticoid response element responds differently to
glucocorticoids in various cells (Pfahl et al.). Furthermore, it has
been shown that the response to inducing glucocorticoids can be modified
by oncogene factors such as those from ras and mos (Jassi et al.). That
these are the specific controls explaining the difference in response
of Fu5-5 cells is not likely however, since in each of these cases one
would predict that the variability in response seen should apply to all
glucocorticoid response element driven genes and not to tyrosine
aminotransferase only. Nevertheless the general principle that the
cellular environment for a promotor is important in influencing its
activity could apply; it would have to be modified to explain how such
influences could operate on only one of several glucocorticoid regulated
genes within a specific cell.

The existing general model for steroid hormone action is inadequate to
explain or predict the behavior of Fu5-5 cells. We expect that a
combination of specific tyrosine aminotransferase gene promotor
structure and its interaction with Fu5-5 regulatory elements will

provide the explanation for the difference in sensitivity of the tyrosine aminotransferase gene in those cells. If this hypothesis proves true, it will extend the general model for glucocorticoid action beyond the simple three component model.

## Post-transcriptional control of tyrosine aminostransferase

In the course of examining the transcription rate response of cells to glucocorticoids, we observed that continued increased transcription of the tyrosine aminotransferase gene alone does not seem to explain the continued induction of tyrosine aminotransferase enzyme in the presence of steroids. The early observations on induction of this gene showed that the steroid hormone acts directly on the responsive cell and is continually required for the continued induction. It was shown that removal of hormone causes a prompt deinduction of enzyme synthesis (Granner et al., 1970). The straightforward transcription model would therefore predict that a sustained level of increased transcription would be necessary to keep message levels elevated. When we followed the time course of transcription activation in HTC and Fu5-5 cells by glucocorticoids, we found that instead of sustained transcription (as measured by nuclear run-off assay) both cells showed what is known as a burst-attenuation response. In constant, maximally inducing concentrations of dexamethasone, the messenger RNA for tyrosine aminotransferase remains elevated for many hours; however, the nuclear run-off assay rises initially, reaches a peak, and promptly declines, almost back to base line levels by 8 hours (table 4). At least two possible explanations could account for these data. One is the iconoclastic view that the nuclear run-off assay does not successfully measure the transcription rate, at later time points becoming invalid. The problem with this interpretation is that its logical extension leads to the invalidation of all nuclear run-off assay data. (If one cannot trust it at a particular time and cannot predict when that time may be, how can one know it is giving reliable data at any time?) Since in several cases there does appear to be a correlation between induction of transcription measured by nuclear run-off and extent of gene product induction (Ringold et al.; Granner et al., 1986), we prefer the more conservative interpretation that the assay can be relied upon to show at least some transcriptional effect when there is one. The data then lead to the conclusion that there is a post-transcriptional component to the induction of the aminotransferase. The messenger RNA is somehow stabilized or its processing improved in the presence of steroid so that the levels of TAT mRNA remain elevated even when the rate of transcription has returned to near-basal levels. Accordingly we carried out induction/deinduction experiments to begin to assess the possibilities. HTC cells were induced to the maximum steady state level, at which point the rate of TAT mRNA transcription appears to be about equal to basal levels (see table 4), and then divided into two parts. In one, steroid was allowed to remain; in the other it was removed by washing the cells, and tyrosine aminotransferase messenger RNA quantities were measured at various time points after. It was found that in the continued presence of the hormone, mRNA levels remained high and constant whereas when steroid was removed, they rapidly fell (figure 4). Unless the withdrawal of steroid represses the rate of TAT mRNA synthesis below the basal level rate, these data suggest that indeed there is a post-transcriptional component (such as TAT mRNA stablization) to the sustained induction of tyrosine aminotransferase in HTC cells by dexamethasone. They do not rule out as yet the possibility that the nuclear run-off assay for unknown reasons becomes unreliable after a few hours in the presence of hormone. However, in view of the fact that other documented instances of post-transcriptional regulation

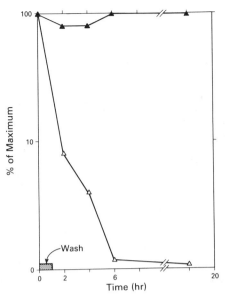

FIGURE 4: Tyrosine aminotransferase mRNA levels in HTC cells after induction and subsequent removal of inducer. A culture of cells was maximally induced by exposure to $10_{-6}$ M dexamethasone overnight. The culture was then divided and half was repeatedly washed in isotonic buffer to remove all inducer ($\triangle$). The other half was returned to steroid-containing medium ($\blacktriangle$). At various subsequent times, samples were removed and assayed quantitatively for TAT mRNA levels.

by steroid hormones have been found, e.g. vitellogenin and albumin (Riegel et al.; Brock and Shapiro), it seems likely that for tyrosine aminotransferase there is a significant component of post-transcriptional regulation as well as transcriptional regulation by glucocorticoids. Obviously the classic model of glucocorticoid transcription induction does not account for these phenomena. It is interesting to note that some of the earliest studies on the mechanism of action of glucocorticoids suggested that there was in fact a post-transcriptional component to their regulation of genes (Thompson et al., 1970). Indeed at one point this model was taken to its logical extreme, when it was proposed that all regulation by glucocorticoids was post-transcriptional (Tomkins et al.). The data upon which this conclusion was based was indirect and relied on inhibitory drug studies, the only methods available at the time. The assumptions required to reach those conclusions were clearly spelled out (Thompson et al., 1970), and subsequently it was found that some of the drugs used were suspect as to the precision of their mechanism of action. Since the side effects of the drugs clouded the interpretation of the data (Steinberg and Ivarie; Kenney), the post-transcriptional model has largely been set aside in the face of the very strong data for transcriptional control that has accumulated in the past decade. Now, however, it seems that it is time to re-examine the older postulates, since in a number of systems it appears that steroids exert both transcriptional and post-transcriptional regulatory controls. There is need to devise a model which will explain all the phenomena. We believe that a further examination of the tyrosine-aminotransferase system will be of value in doing so.

REFERENCES

Baxter, J.D. and Rousseau, G.G. Glucocorticoid Receptors. In: Glucocorticoid Hormone Action. Baxter, J.D. and Rousseau, G.G. (Eds.), Springer-Verlag, Berlin, Heidelberg, New York, pp. 62-63 (1979).

Baxter, J.D. and Tomkins, G.M. The relationship between glucocorticoid binding and tyrosine aminotransferase induction in hepatoma tissue culture cells. Proc. Natl. Acad. Sci. USA 65:709-715 (1970).

Baxter, J.D. and Tomkins, G.M. Specific cytoplasmic glucocorticoid hormone receptors in hepatoma tissue culture cells. Proc. Natl. Acad. Sci. USA 68:932-937 (1971).

Brock, M.L. and Shapiro, D.J. Estrogen stabilizes vitellogenin mRNA against cytoplasmic degradation. Cell 34:207-214 (1983).

Fischer, J.A. and Maniatis, T. Drosophila Adh: a promoter element expands the tissue specificity of an enhancer. Cell 53:451-461 (1988).

Gadson, P.G., Simons, S.S., Jr. and Thompson, E.B. Differential induction of tyrosine aminotransferase gene transcription of glucocorticoids in hepatoma cell lines. (Submitted for publication).

Granner, D.K., Sasaki, K. and Chu, D. Multihormonal regulation of phosphoenolpyruvate carboxykinase gene transcription, Ann. NY Acad. 478:175-190 (1986).

Granner, D.K., Thompson, E.B. and Tomkins, G.M. Dexamethasone phosphate-induced synthesis of tyrosine aminotransferase in hepatoma tissue culture cells. J. Biol. Chem. 245:1471-1478 (1970).

Greengard, O. and Acs, G. The effect of actinomycin on the substrate and hormonal induction of liver enzymes. Biochim. Biophys. Acta, 61:652-653 (1962).

Jantzen, H.M., Strahle, U., Gloss, B., Stewart, F., Schmid, W., Boshart, M., Miksicek, R. and Schütz, G. Cooperativity of glucoccorticoid response elements located far upstream of the tyrosine aminotransferase gene. Cell 49:29-38 (1987).

Jassi, R., Salmons, B., Muellenger, D. and Groner, B. The v-mos and h-ras oncogene expression represses glucocorticoid hormone-dependent transcription from the mouse mammary tumor virus LTR. EMBO J. 5:2609-2616 (1986).

Kenney, F.T. Hormonal regulation of synthesis of liver enzymes. In: Mammalian Protein Metabolism, Munro, H.T. (Ed.), Academic Press, New York, pp. 131-150 (1970).

Lee, F., Mulligan, R., Berg, P. and Ringold, G. Glucocorticoids regulate expression of dihydrofolate reductase cDNA in mouse mammary tumor virus chimeric plasmids, Nature 294:228-232 (1981).

Lin, E.C.C. and Knox, W.E. Adaptation of the rat liver tyrosine-α-ketoglutarate transaminase. Biochim. Biophys. Acta. 26:85-88 (1957).

Lin, E.C.C. and Knox, W.E. Specificity of the adaptive response of tyrosine-α-ketoglutarate transaminase in the rat. J. Biol. Chem. 233:1186-1189 (1958).

Loeb, J.N. and Strickland, S. Hormone binding and coupled response relationships in systems dependent on the generation of secondary mediators. Mol. Endo. 1:75-82 (1987).

Mercier, L., Thompson, E.B. and Simons, S.S., Jr. Dissociation of steroid binding to receptors and steroid induction of biological activity in a glucocorticoid-responsive cell. Endocrinol. 112:601-609 (1983).

Miller, P.A. and Simons, S. S., Jr. Comparison of glucocorticoid receptors in two rat hepatoma cell lines with different sensitivities to glucocorticoids and antiglucocorticoids. Endocrinol, 122:2990-2998 (1988).

Pfahl, M., Payne, J., Benbrook, D., and Wu, K.C. Differential activation of a hormone-responsive element in various cell lines. In: Steroid Hormone Action (Ed: G. Ringold), Alan R. Liss, New York, pp. 161-168 (1988).

Riegel, A. T., Aitken, S.C., Martin, M.B. and Schoenberg, D.R. Posttranscriptional regulation of albumin gene expression in Xenopus liver: evidence for an estrogen receptor-dependent mechanism. Mol. Endo. 1:160-167 (1987).

Ringold, G.M. Steroid hormone regulation of gene expression. Ann. Rev. Pharmacol. Toxicol. 25:529-566 (1985).

Rousseau, G.G., Baxter, J.D. and Tomkins, G.M. Glucocorticoid receptors: relations between steroid binding and biological effects. J. Mol. Biol. 67:99-115 (1972).

Scherer, G., Schmid, W. Strange, C.M. Rowckamp, W. and Schütz, G. Isolation of cDNA clones coding for rat tyrosine aminotransferase. Proc. Natl. Acad. Sci. 79:7205-7208 (1982).

Schmid, W., Jantzen, M., Mayer, D., Jastorff, B. and Schütz, G. Transcription activation of the tyrosine aminotransferase gene by glucocorticoids and cAMP in primary hepatocytes. J. Biochem. 165:499-506 (1987).

Schmid, W., Muller, G., Schütz, G. and Gluecksohn-Waelsch, S. Deletions near the albino locus on chromosome 7 of the mouse affect the level of tyrosine aminotransferase mRNA. Proc. Natl. Acad. Sci. 82:2866-2869 (1985).

Schöler, H.R. and Gruss, P. Specific interaction between enhancer-containing molecules and cellular components. Cell 36:403-411 (1984).

Simons, S.S., Jr., Miller, P.A., Wasner, G., Miller, N.R. and Mercier, L. Inverse correlation between dexamethasone 21-mesylate agonist activity and sensitivity to dexamethasone for induction of tyrosine aminotransferase in rat hepatoma cells. J. Steroid Biochem. 31:1-7 (1988a).

Simons, S.S., Jr. Mercier, L., Miller, N.R., Miller, P.A., Oshima, H., Sistare, F.D., Thompson, E.B. and Wasner, G. Differential modulation of gene induction by glucocorticoids and antiglucocorticoids in rat hepatoma tissue culture cells. In Press: Cancer Research (1988b).

Steinberg, R.A. and Ivarie, R.D. Posttranscriptional regulation of glucocorticoid-regulated functions. In: Glucocorticoid Hormone Action. (Eds. Baxter, J.D. and Rousseau, G.G.), Springer-Verlag, Berlin, Heidelberg, New York, pp. 291-302 (1979).

Strickland, S. and Loeb, J.N. Obligatory separation of hormone binding and biological response curves in systems dependent upon secondary mediators of hormone action. Proc. Natl. Acad. Sci. USA 78:1366-1370 (1981).

Strobl, J.S., Dannies, P.S. and Thompson, E.B. Somatic cell hybridization of growth hormone-producing rat pituitary cells and mouse fibroblasts results in extinction of growth hormone expression via a defect in growth hormone RNA production. J. Biol. Chem. 257:6588-6594 (1982).

Thompson, E.B. Tomkins, G.M. and Curran, J.F. Induction of tyrosine alpha-ketoglutarate transaminase by steroid hormones in a newly established tissue culture cell line. Proc. Natl. Acad. Sci. USA 56:296-303 (1966).

Thompson, E.B., Granner, D.K., and Tomkins, G.M. Superinduction of tyrosine aminotransferase by actinomycin D in rat hepatoma (HTC) cells. J. Mol. Biol. 54:159-175 (1970).

Tomkins, G.M., Gelehrter, T.D., Granner, D.K., Martin, D., Jr., Samuels,

H.H. and Thompson, E.B. Control of specific gene expression in higher organisms. Science 166:1474-1480 (1969).

Wasner, G. and Simons, S.S., Jr. Differential sensitivity of HTC and Fu5-5 cells for induction of tyrosine aminotransferase by 3',5'-cyclic adenosine monophosphate. Mol. Endo. 1:109-120 (1987).

Wasner, G., Oshima, H., Thompson, E.B., and Simons, S., Jr. Unlinked regulation of the sensitivity of primary glucocorticoid-inducible responses in transfected Fu5-5 rat hepatoma cells. Mol. Endo. 2:1009-1017 (1988).

Weiss, M.C. and Chaplain, M. Expression of differentiated functions in hepatoma cell hybrids: reappearance of tyrosine aminotransferase inducibility after the loss of chromosomes. Proc. Natl. Acad. Sci. USA 68:3026-3030 (1971).

DISCUSSION OF THE PAPER PRESENTED BY E.B. THOMPSON

STANCEL: You showed data on the decrease in myc RNA. Do you have
any idea about the levels and half-life of the protein product in
these cells?

THOMPSON: We have not as yet made any measurements of myc-specific
protein. Of course it is our intention to do so. In most systems,
myc mRNA and protein both have rather short half-lives.

ROY: You showed that GC inhibits cell division and c-myc is down-
regulated after cells reach confluency. It raises the question whether
the effect of GC on c-myc is a direct one or mediated indirectly
through some cycloheximide-sensitive steps?

THOMPSON: Although glucocorticoids down-regulated c-myc mRNA in CEM
lymphoid cells, we do not know the mechanism. In P1798 cells, Aubrey
Thompson has data suggesting glucocorticoids down-regulate myc at the
transcriptional level (Molec. Endo 1:899, 1987), but the regulation of
this gene in various systems is complex, involving nearly all possible
levels. Therefore we will have to explore the questions directly in
CEM cells and not rely on analogy with other cell systems.

SCHMID: I want to make a comment pertaining to the first part of your
talk. HTC cells lack a DNAase hypersensitive region 3.6 kb upstream
as detected by work done by Francis Stewart in our lab. This site,
as detected by Michael Beshard in our lab, is a strong cell specific
enhancer which also seems to be involved in mediating extinction.

THOMPSON: Thank you for the information. Has your data been published?


DISCUSSANTS: G. STANCEL, E.B. THOMPSON, A.K. ROY AND W. SCHMID

# STEROID RESPONSE ELEMENTS. DEFINITION OF A MINIMAL PROMOTER AND INTERACTION WITH OTHER ACTIVATING SEQUENCES

W. SCHMID, U. STRAHLE, R. MESTRIL, G. KLOCK, W. ANKENBAUER, and G. SCHUTZ, Institute of Cell and Tumor Biology, German Cancer Research Center, Heidelberg.

Control of gene expression by steroid hormones is exerted mainly by modulation of the transcriptional activity of the regulated genes. The response to the various steroid hormones is highly specific. It is mediated by binding of the hormone to its specific receptor which, after binding, attains the capacity to recognize a specific DNA region on regulated genes. This receptor-DNA interaction eventually leads in a still-unknown way to an increase of the rate of transcription of the target gene (Yamamoto, 1985).

Sequences mediating glucocorticoid and estrogen responsiveness have been identified using DNA transfection techniques in several steroid-regulated genes (Yamamoto, 1985; Ringold, 1985; Scheidereit et al, 1986; Jantzen et al, 1987; Klein-Hitpass et al, 1986, 1988; Klock et al, 1987; Martinez et al, 1987; Ankenbauer et al, 1988). In many cases, binding of purified receptor to a specific DNA sequence was demonstrated by footprinting techniques (Payvar et al, 1983; Scheidereit et al, 1983; Karin et al, 1984; Miksicek et al, 1986; Danesch et al, 1987; Jantzen et al, 1987). It could be shown that the sequences required for induction of transcription are coincident with binding sites of the respective receptors. Evidence for a hormone-triggered interaction in intact cells of the glucocorticoid receptor with its target sequence on the tyrosine aminotransferase (TAT) gene was demonstrated by genomic sequencing techniques, indicating that the interaction in vivo was strictly dependent on the presence of hormone and restricted to cells expressing TAT (Becker et al, 1986).

The recognition sequences of glucocorticoid and estrogen receptors constitute a family of closely related sequences (Klock et al, 1987; Strahle et al, 1987; Martinez et al, 1987; Ankenbauer et al, 1988; Klein-Hitpass et al, 1988). All of these elements are partial or perfect palindromes (Strahle et al, 1987; Klock et al, 1987; Martinez et al, 1987) of 15 base pairs length. Moreover, base substitutions in either halfpart of the palindrome affect inducibility, suggesting a dimeric structure of the receptor protein binding to the recognition sequence (Strahle et al, 1987).

Several observations suggest that receptor action is dependent on interaction with other transcription factors. For example, glucocorticoid induction of the murine mammary tumor virus (MMTV) requires the presence of an intact nuclear factor I (NF I) site (Kuhnel et al, 1986; Cordingley et al, 1987; Miksicek et al, 1987). Similarly, glucocorticoid induction of the tryptophan oxygenase (TO) gene is dependent on the integrity of a CACCC box which was first described as a promoter element of the β-globin gene (Myers et al, 1986), suggesting that a transcription factor binding to that sequence is required for hormonal induction of this gene (Danesch et al, 1987; Schule et al, 1988).

As it was previously shown (Strahle et al, 1987; Klock et al, 1987) that the 15 bp GRE as well as the 15 bp estrogen response element (ERE) is active in front of the well-characterized herpes simplex virus TK promoter (McKnight et al, 1981, 1982), we tried to reduce the complexity of this system further, asking the question whether a GRE or a ERE alone would be sufficient for hormonal induction or whether more complex structures are required to build up a hormone-dependent enhancer (Yamamoto, 1985).

**Methods:**

All constructs are derivatives of pBLCAT 2 (Luckow and Schutz, 1987) or pTATCAT (Jantzen et al, 1987). Purification and cloning of the synthetic oligonucleotides were done by standard procedures (Maniatis et al, 1982).

Cell culture was done as described elsewhere (Jantzen et al, 1987), using a clonal isolate (FTO2B-3) of FT02B hepatoma cells which has high levels of TAT. Ltk-cells were transfected by DEAE-dextran as described (Jantzen et al, 1987); MCF7, XC and FT02B-3 cells by electroporation as described elsewhere (Strahle et al, 1988).

**RESULTS:**

**The Response Elements for Glucocorticoids, Gestagens, Estrogens, Androgens and Ecdysterone Are Highly Conserved.**

Analysis of the sequences recognized by glucocorticoid, progesterone and estrogen receptors has shown that these elements are closely related. In fact, single base pair mutations of the 15 base pair palindromic motif show the same detrimental effect when tested for glucocorticoid and gestagen induction in cells expressing either glucocorticoid or progesterone receptor (Strahle et al, 1987). An ERE might differ from a GRE by a single symmetrical base pair exchange (Klock et al, 1987). Notably, an ERE can also be recognized by the T3 receptor (Glass et al, 1988). It has previously been reported that the MMTV LTR is responsive, not only to glucocorticoids and gestagen (Cato et al, 1986), but also to a lesser extent to androgens as well (Cato et al, 1988). Therefore, constructs containing the TAT GRE which mediates glucocorticoid as well as gestagen inducibility to the TK promoter were also analyzed for androgen response. As seen in Figure 1, these sequences also permit a five-fold induction by dihydrotestostdrone in cells possessing androgen receptors.

The expression of the small heat shock genes of Drosophila melanogaster is controlled by ecdysterone. Deletion analysis has defined fragments conferring ecdysterone inducibility in the 5'-flanking region of hsp 23 and hsp 27 (Mestril et al, 1986; Riddihough and Pelham, 1987). The hsp 27 promoter contains 15 bp sequences which show striking homology to the steroid response elements of vertebrates. Therefore, a 15 bp sequence from hsp 27 was positioned upstream of a hsp 70 promoter - CAT construct truncated at position -50 of the hsp 70 promoter and the resulting construct was tested after transfection in Schneider cells for inducibility by β-ecdysterone. As shown in Figure 1, the construct was about seven-fold inducible by the insect steroid hormone; duplication of the element results in a more than 100-fold inducibility.

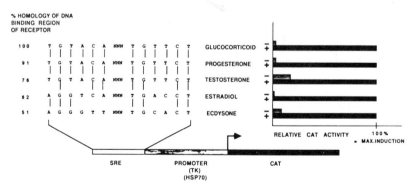

**FIGURE 1.** The response elements for glucocorticoids, testosterone, progesterone, estradiol and ecdysone are closely related.

The response elements when inserted upstream of the thymidine kinase promoter of herpes simplex (TK) or, in case of the ecdysone response element, upstream of the truncated promoter of the heat shock protein 70 gene of Drosophila melanogaster, mediate induction of CAT activity by the respective steroid after transfection into responsive cell lines. Identical bases in steroid response elements are indicated by lines (numbers indicate % of homology to the DNA binding domain of the glucocorticoid receptor, Evans, 1988)

In addition to the effects of the inducers on CAT expression, the homologies of the recognition sequences, as well as the degree of conservation of the DNA-binding region of the receptors which interact with these sequences, are shown. Our results suggest that the strongly conserved DNA-binding regions of the different steroid receptors might have co-evolved with the DNA sequences with which they interact. The finding that the same sequence is able to interact with different receptors implies that the high specificity of hormonal action operates on additional levels of control. One level is certainly the absence or presence of receptor in a given cell, but at least in the case of glucocorticoid receptor which is found in most of the cells other mechanisms, like interaction with other limiting transcription factors, might be operative as well.

**A Steroid Hormone Response Element is Active in Front of a TATA Box but Not from a Far Upstream Position.**

The experiments summarized in Figure 1 clearly indicate that steroid receptors are active when linked to the TK promoter. As there was circumstantial evidence that hormone induction needs cooperation of the receptor with additional transcription factors as already mentioned in the Introduction, we reduced the complexity of the TK promoter by placing a GRE or an ERE directly upstream of the TATA box, thus deleting both distal elements of the TK promoter (McKnight et al, 1984). As it was shown that two GREs might act synergistically (Jantzen et al, 1987), constructs containing multiple GREs were also tested. As seen in Figure 2A, a single GRE or ERE upstream of a TATA box is clearly capable of conferring hormonal inducibility. Duplication of the elements has a much larger than additive effect (figure 2B) that is not substantially increased by further multimerization. Equivalent constructs containing the TATA box of the hsp70 promoter are equally effective (data not shown).

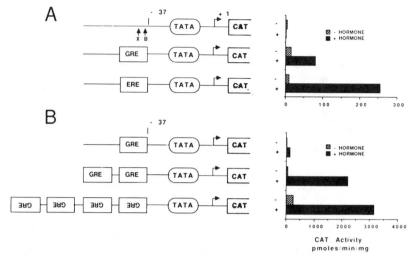

**FIGURE 2. A glucocorticoid response element (GRE) upstream of a TATA motif is sufficient for induction.** One or two copies of a synthetic oligonucleotide containing the 15 bp long GRE, TGTACAGGATGTTCT or a ERE, AGGTCACAGTGACCT, were inserted upstream of the TATA box of the TK promoter. After transfection into MCF-7 cells, CAT activity was determined in the absence or presence of hormones.

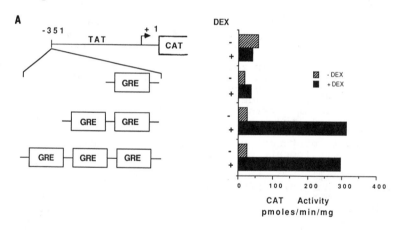

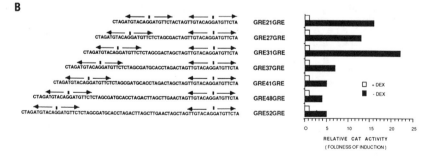

A single 15 bp GRE, when placed at position - 351 of
the TATA promoter, showed no or only very minor
inducibility, depending on the cell type used for
testing (figure 3A). Dimerization again leads to a
very potent element. Trimerization does not result in
a stronger element. To see whether the synergism of
two GREs requires a certain topological alignment, as
observed for the interaction between enhancer and
promoter region in SV 40 DNA (Takahashi et al, 1986),
the center-to-center distance between the two GREs was
varied by increments of 5 bp. As seen in figure 3B,
synergism is maintained within a distance between 22
and 31 bp without significant change. Exceeding that
distance, the synergism is less pronounced. These
results suggest that no strict stereo alignment is
required to achieve synergism. Glucocorticoid and
estrogen receptors can also interact synergistically,
as shown by Ankenbauer et al (1988). Recent
experiments suggest that this synergism of two adjacent
GREs is brought about by cooperative binding of
receptor molecules to two neighboring hormone response
elements (W. Schmid, unpublished).

**Several Transcription Factors Cooperate With The
Glucocorticoid Receptor In a Cell-Type-Specific Manner**

The finding that a single GRE is almost inactive at
position - 351 of the TAT promoter was unexpected, as a
previous construct which contained the identical 15 bp
GRE plus additional flanquing sequences of the
glucocorticoid inducible enhancer of the TAT gene was
strongly inducible at this position as well as 2 kb
further upstream (Jantzen et al, 1987). A close
inspection of the 35 bp fragment used in this construct
revealed the presence of a CCAAT motif 6 bp upstream of
the GRE (see figure 4, bottom). As seen in Figure 4,
deletion of mutation of the CCAAT motif abolishes
inducibility by glucocorticoids.

**FIGURE 3 A. A single glucocorticoid response element
is not sufficient for activation from an upstream
position.** One, two or three copies of the 15 bp long
GRE (TGTACAGGATGTTCT) were cloned upstream of the TAT
promoter driving the expression of the gene. The
constructs were transfected into Ltk cells and hormone
inducibility was determined by comparing CAT activity
in extracts prepared from uninduced or induced cells.
**FIGURE 3 B.** The distance between two GREs was varied
by increments of about 5 base pairs. The effect of
distance variation was measured by determination of CAT
activity.

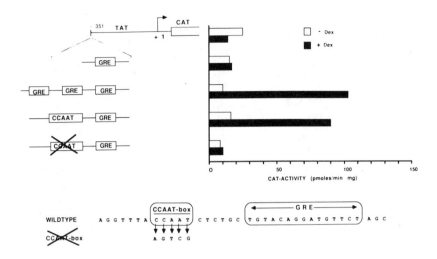

**FIGURE 4. The CCAAT-box binding protein cooperates with the glucocorticoid receptor.**
A single GRE, when inserted 351 bp upstream of the tyrosine aminotransferase promoter (TAT), is not sufficient for induction. However, a short fragment at the same position which comprises the sequences between -2527 bp and -2492 bp of the glucocorticoid inducible enhancer of the tyrosine aminotransferase gene and contains a CCAAT-box in close proximity to a single GRE (sequence is depicted at the bottom) leads to a strong dexamethasone-dependent increase in CAT activity. Mutation of the CCAAT-box by a cluster of transversions abolishes inducibility by dexamethasone (DEX). Constructs were transfected into Ltk cells.

This observation prompted us to investigate whether other transcription factors might interact with the glucocorticoid receptor in a similar fashion. To address this question, the 15 bp GRE was linked to either the NF I recognition sequence of the LTR of MMTV (Miksicek et al, 1987), to the CACCC element of the β-globin promoter (Myers et al, 1986) or to the SP1 recognition sequence from the second distal element of the TK promoter (McKnight et al, 1984) and placed at position -351 of the TATA promoter. These constructs were then tested for hormone inducibility in the different cell lines shown in Figure 5. The results demonstrated that each combination creates an active element but activity differs in the cell lines tested. The strongest combination was the CACCC-GRE construct when tested in hepatoma cells. This combination was less effective in the other cell lines tested. The other combinations showed a different degree and spectrum of hormone inducibility.

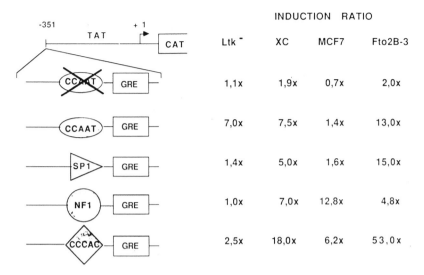

**FIGURE 5. Transcription factors cooperate with the glucocorticoid receptor in a cell-specific manner.** TAT-CAT fusion genes carrying various combinations of a single GRE with transcription factor binding sites upstream of the TAT promoter were transfected into Ltk-, XC, MCF-7 and FTO2B-3 cells and analyzed for inducibility of CAT expression by dexamethasone.

Taken together, these results show that several transcription factors can act synergistically with the glucocorticoid receptor. Most probably, the activity or abundance of these factors differs in various cell types. In native hormonally-controlled enhancers, these regulatory proteins can be constitutive parts of a hormone - dependent enhancer complex as described for the MMTV LTR (Kuhnel et al, 1986), the MSV LTR (Miksicek et al, 1987) or the TO gene (Danesch et al, 1987). Such combinations of different transcription factor binding sites with hormone response elements to form an inducible enhancer allow modulation of the activity of the receptor, leading to a more selective mode of hormone action.

## CONCLUSIONS:

The data presented here demonstrate that steroid hormone response elements form a family of related sequences. Some hormone receptors are able to bind to an identical sequence despite their very distinct biological activities. When positioned close to the TATA box, a single recognition sequence is sufficient for hormone-dependent transcriptional activation. To create a position-independent element, receptors have to interact with the same or a different receptor species or with other transcription factors like the proteins interacting with a CCAAT-, a CACCC, a SP 1 box or a NF I recognition sequence.

**Acknowledgments:**

We thank W. Fleischer for synthesis of the oligonucleotides and Ms. C. Schneider for excellent secretarial assistance. We thank the Fonds der Chemischen Industrie for financial support.

**References:**

Ankenbauer, W., Strahle, U., and Schutz, G. (1988) Proc. Natl. Acad. Sci. USA, in press.

Becker, P.B., Gloss, B., Schmid, W., Strahle, U., and Schutz, G. (1986) Nature 324:686-688.

Cato, A.C.B., Miksicek, R., Schutz, G., Arnemann, J., and Beato, M. (1986) EMBO J. 5:2237-2240.

Cato, A.C.B., Skroch, P., Weinmann, J., Butkeraitis, P., and Ponta, H. (1988) EMBO J. 7:1403-1410.

Cordingley, M.G., Rieget, A.T., and Chambon, P. (1987) Cell 48: 261-270.

Danesch, U., Gloss, B., Schmid, W., Schutz, G., Schule, R. and Renkawitz, R. (1987) EMBO J. 6:625-630.

Glass, C.K., Holloway, J.M., Devary, O.V., and Rosenfeld, M.G. (1988) Cell 54:313-323.

Jantzen, H.M., Strahle, U., Gloss, B., Stewart, F., Schmid, W., Boshart, M., Miksicek, R. and Schutz, G. (1987) Cell 49:29-38.

Karin, M., Haslinger, A., Holtgreve, A., Richards, R.I., Krauter, P., Westphal, H.M. and Beato, M. (1984) Nature 308: 513-519.

Klein-Hitpass, L., Schorpp, M., Wagner, U., and Ryffel, G. (1986) Cell 46:1053-1061.

Klein-Hitpass, L., Ryffel, G., Heitlinger, E., and Cato, A.C.B. (1988) Nucleic Acid Res. 16:647-663.

Klock, G., Strahle, U., and Schutz, G. (1987) Nature 329: 734-736.

Kuhnel, B., Buetti, E., and Diggelmann, H. (1986) J. Mol. Biol. 190:367-378.

Luckow, B., and Schutz, G. (1987) Nucleic Acid Res. 15: 5490.

Maniatis, T., Fritsch, E.F., Sambrook, J. (1982) Molecular cloning: A laboratory manual. Cold Spring Harbor, New York.

Martinez, E., Givel, F., and Wahli, W. (1987) EMBO J. 6: 3179-3727.

McKnight, S.L., Gavis, E.R., Kingsbury, R.C., and Axel, R. (1981) Cell 25:385-398.

McKnight, S.L. (1982) Cell 31:355-365.

McKnight, S.L., Kingsbury, R.C., Spence, A., and Smith, M. (1984) Cell 37:253-262.

Mestril, R., Schiller, P., Amin, J., Klapper, J., and Voellmy, R. (1986) EMBO J. 5:1667-1673.

Miksicek, R., Heber, A., Schmid, W., Danesch, U., Posseckert, G., Beato, M., and Schutz, G. (1986) Cell 46: 283-290.

Miksicek, R., Borgmeyer, U., and Nowock,I. (1987) EMBO J. 6: 1355-1360.

Myers, R.M., Tilly, K., and Maniatis, T. (1986) Science 232: 613-618.

Payvar, F., DeFranco, D., Firestone, G.L., Edgar, B., Wrange, O., Okret, S., Gustafson, J.A., and Yamamoto, K.R. (1983) Cell 351:381-392.

Riddihough, G., and Pelham, H.R.B. (1986) EMBO J. 5: 1653-1658.

Riddihough, G., and Pelham, H.R.B. (1987) EMBO J. 7: 3729-3734.

Ringold, G.A. (1985) Ann. Rev. Pharmacol. Toxicol. 25: 529-566.

Scheidereit, G., Geisse, S., Westphal, H.M., and Beato, M. (1983) Nature 304:749-752.

Scheidereit, G., Westphal, H.M., Carlson, C., Bosshard, H., and Beato, M. (1986) DNA 5:383-391.

Schule, R., Muller, M., Otsuka-Murakami, H., and Renkawitz, R. (1988) Nature 332:87-90.

Strahle, U., Klock, G., and Schutz, G. (1987) Proc. Natl. Acad. Sci. USA 84:7871-7875.

Strahle, U., Schmid, W., and Schutz, G. (1988) EMBO J. (in press).

Takahashi, K., Vigneron, M., Matthes, H., Wildeman, A., Zenke, M., and Chambon, P. (1986) Nature 319:121-126.

Yamamoto, K.R. (1985) Ann. Rev. Genet. 19:209-252.

DISCUSSION OF THE PAPER PRESENTED BY W. SCHMID

THOMPSON: In the GRE 1 linked vs. unlinked competition experiment, did you try, as a control, unlinked GRE flanked by DNA as long as the total length of DNA in the linked GRE constructs?

SCHMID: No, we did not do that so far, but the length of the resulting oligonucleotides is about 40 base pairs so that the GRE containing fragments should be stable under the conditions of the shift assay.

DISCUSSANTS: E.B. THOMPSON AND W. SCHMID.

GHF-1, A TISSUE-SPECIFIC TRANSCRIPTION FACTOR, IS A HOMEOBOX PROTEIN

José-Luis Castrillo[*], Lars E. Theill, Mordechai Bodner and Michael Karin

Department of Pharmacology. School of Medicine. M-036.
University of California, San Diego. La Jolla, CA-92093.

Introduction

The growth hormone (GH) gene is uniquely expressed in specialized cells, known as somatotropes, in the anterior pituitary. Previous analysis has shown that the difference between GH gene activity in somatotropes and hepatocytes exceeds $10^8$-fold (Ivarie et al, 1983). To achieve this level of control, a highly specific mechanism should restrict GH expression to the correct cell type. The GH-expressing somatotrope is derived from an acidophylic stem cell that appears during the ontogeny of the anterior pituitary which also gives rise to the prolactin (PRL)-expressing lactotrope (Chatelain et al., 1979). It is generally accepted that during development and differentiation of the rat pituitary, the acidophylic stem cell first gives rise to somatotropes ($GH^+$, $PRL^-$), which in turn give rise to additional somatotropes and to mammosomatotropes ($GH^+$, $PRL^+$). These latter cells are the major precursors of lactotropes ($GH^-$, $PRL^+$), the last cell type to appear along this lineage (Hoeffler et al., 1985; Behringer et al., 1988). Various data indicate the existence of a subpopulation of $GH^+$ cells which can switch to $PRL^+$ cells and $PRL^+$ cells which can switch to $GH^+$ cells, under the influence of certain hormones such as estrogens and GH-releasing factor (Leong et al., 1985; Strattmann et al., 1974).

The regulation of cell type differentiation in this progressive lineage offers an interesting experimental system for studying the control of cellular differentiation.

Previous work in this laboratory indicated that the specific expression of the GH gene in GH-expressing pituitary tumor cells is dictated by a pituitary specific promoter region containing two binding sites for a single transcription factor, GHF-1 (Lefevre et al., 1987). These tumor cell lines also express PRL and therefore are equivalent to the mammosomatotropes described above (Tashjian et al., 1986; Bancroft, 1981). As determined by DNA-binding assays, GHF-1 is present only in GH-expressing cell lines (Lefevre et al., 1987) and it disappears in stable pituitary x fibroblast somatic cell hybrids in which GH-expression is extinguished (McCormick et al., 1988). Addition of GHF-1 to extracts of non-expressing cells leads to the activation of GH transcription (Bodner and Karin, 1987). To understand the mechanisms that control the expression of GHF-1 itself and how it activates transcription of the GH gene, we have purified the protein to near homogeneity and determined a partial amino acid sequence of a GHF-1 derived peptide. Screening of bovine pituitary and rat GC cell cDNA libraries, with synthetic oligonucleotide probes that correspond to the GHF-1 peptide sequence, resulted in isolation of cDNA clones encompassing the complete coding region of bovine and rat GHF-1. Sequence analysis revealed that a region of the GHF-1 protein, near its C-terminus, exhibits a high degree of sequence homology to the homeobox, a conserved sequence motif previously identified in genes that regulate development of the fruit-fly D. melanogaster and the yeast S. cerevisiae (see Scott and Caroll, 1987; Gehring, 1987 for reviews). This highly conserved structure was also found in many different taxa of the animal kingdom, including mammals (see Gehring, 1987 for review). The function of homeobox containing proteins in taxa other than Drosophila is not clear. However, because of their similarity to the Drosophila homeotic genes, homeobox containing genes in other systems are likely to encode factors that regulate development.

GHF-1 is the first mammalian homeobox containing gene shown to function as a transcription factor and is also the first homeobox containing gene from any animal species whose target gene is known. These findings, when taken together with the cell-type specificity of GH expression, suggest that GHF-1 plays a major role in regulating the differentiation of GH-expressing cells in the anterior pituitary.

Materials and Methods

Isolation of GHF-1 cDNA clones

Two oligodeoxynucleotide hybridization probes were synthesized according to the partial amino acid sequence of GHF-1 (Bodner et al. 1988). Two different libraries were used: A bovine pituitary cDNA library in λgt10 was kindly provided by Dr. G. Bell (University of Chicago) and a rat GC cell cDNA library in λgt11 was constructed according to standard procedures (Maniatis et al., 1982). Several independent clones hybridizing to both probes were isolated from each library at an approximate frequency of 1 to 3,000. The sequences of the longest cDNA clones from each source were determined by double stranded dideoxy sequencing (Hattori and Sakaki, 1986) of both strands.

Development of Anti-GHF-1 Antibodies

The MB1 peptide obtained according to partial sequencew of GHF-1 (Bodner

et al., 1988) was coupled to bovine serum albumin using a water soluble
carbodiimide (Mosbach et al., 1972) .200 $\mu$g of conjugates were injected
into New Zealand White rabbits. The injection was repeated every 3
weeks. After the 3$^{rd}$ injection, the rabbits were bled and their sera
were tested for reaction with the MB1 peptide spotted onto nitrocellulose filters by a solid-state enzyme linked immunoassay (ELISA;
Engvall, 1980). This antiserum (anti-GHF-1) was affinity purified on a
column of MB1 peptide immobilized on CH-Sepharose 4B (Pharmacia) using
water soluble carbodiimide mediated coupling (Mosbach et al., 1972).
Antibodies were eluted with 0.2 M glycine/HCl (pH 2.5) and immediately
neutralized by the addition of 2 M Tris/HCl (pH 8.4). Eluted fractions
were tested for reaction with purified GHF-1 and the synthetic peptides
were tested by ELISA and Western blotting using GAR-G-gold and silver
enhancement kits (BioRad). The fractions showing the highest level of
reactivity were pooled and used in all further experiments. Anti-rat
growth hormone (anti-GH) was developed by Dr. Yagha Sinha (Whittier
Institute, La Jolla). Secondary antibodies were purchased from Copper
Biomedicals.

Immunofluorescence of Cultured Cells

GC and HeLa cells were grown on sterile cover slides for 3 days at 37$^{\circ}$C.
The cells were briefly washed in PBS (137 mM NaCl, 2.7 mM KCl, 8 mM Na$_2$
HPO$_4$ and 8 mM KH$_2$ PO$_4$; pH 7.5) and fixed as described (Bodner et al.
1988). The cells were incubated with specific antisera (anti-GHF-1
(1/200) or anti-GH (1/500)) for 1 hr. and washed five times in PBS plus
5 mM glycine with 2% NGS. The cells were then incubated in goat
anti-rabbit rhodamine (1/200) or goat anti-monkey fluorescein (1/200)
for 20 min. and washed extensively in PBS.

Experimental

Isolation of GHF-1 cDNA Clones

GHF-1 was purified to homogeneity from extracts of GC cells, a GH-expressing cell line derived from a rat anterior pituitary tumor (Tashjian et al., 1968), by SDS-polyacrylamide gel electrophoresis (SDS-PAGE) and DNA-specific affinity chromatography method (figure 1). The purified protein had a relative molecular mass of 33K (Bodner et al., submitted). Digestion of 5 μg of purified GHF-1 with V-8 protease generated two peptides whose sizes are 19K and 14K. After separation by SDS-PAGE, these peptides were transferred to Immobilon membranes and identified by staining. Membrane strips, containing each of the peptides, were excised and subjected to amino acid sequence analysis (Matsudaira, 1987). The 19K polypeptide yielded a sequence of 23 amino acids. Two

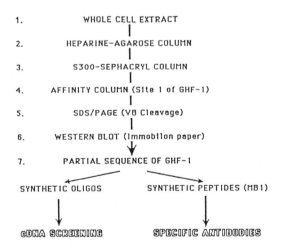

1.     WHOLE CELL EXTRACT

2.     HEPARINE–AGAROSE COLUMN

3.     S300–SEPHACRYL COLUMN

4.     AFFINITY COLUMN (Site 1 of GHF–1)

5.     SDS/PAGE (V8 Cleavage)

6.     WESTERN BLOT (Immobilon paper)

7.     PARTIAL SEQUENCE OF GHF–1

SYNTHETIC OLIGOS          SYNTHETIC PEPTIDES (MB1)

cDNA SCREENING          SPECIFIC ANTIBODIES

Figure 1: Purification of Growth-hormone factor 1 (GHF-1)

different deoxyoligonucleotide hybridization probes were synthesized according to this amino acid sequence and used for screening of bovine pituitary and rat GC cell cDNA libraries (Bodner et al. 1988). Several cDNA clones that hybridize to both probes were isolated from each of the libraries(figure 2). The rat ORF contains a sequence that is identical to the partial amino acid sequence of the GHF-1 19K peptide (Bodner et al. 1988).

Immunofluorescent detection of GHF-1

To prove that the 19K peptide whose partial amino acid sequence was determined by protein sequencing is indeed derived from GHF-1, an oligopeptide was synthesized according to that sequence. The peptide,

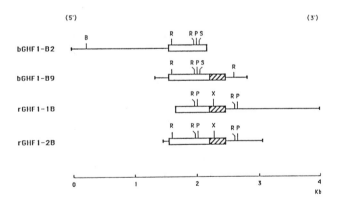

Figure 2: <u>Clones of bovine and rat GHF-1 (bGHF-1, rGHF-1)</u>
Boxes and lines indicate ORF and untranslated regions, respectively. Hatched boxes indicate the homeobox domain. B, Bam H1; P, Pst 1; R, Eco R1; S, Sst1.

MB1, was coupled to BSA and injected into rabbits (figure 3). The antisera was affinity purified and checked by Western blot and no significant reaction with other cellular protein was found (McCormick et al. 1988). In addition, the antibody specifically reacted with the DNA binding activity originally defined as GHF-1 (Bodner et al. 1988). GH-expressing GC cells and GH-nonexpressing HeLa cells, grown on glass cover-slips, were fixed and first incubated with either anti-GH or anti-GHF-1 antibodies and then with rhodamine or fluorescein labelled secondary antibodies.As shown in figure 4, incubation with anti-GHF-1 results in staining of the nuclei of GC cells, but not of HeLa cells.

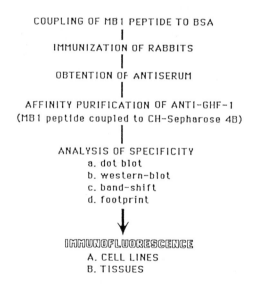

Figure 3: <u>Purification of specific-peptide antibody to GHF-1</u>

Nuclear localization is consistent with the site of action of GHF-1 since it is a transcription factor. Incubation with anti-GH resulted in staining of cytoplasmic granules in GC, but not in HeLa cells. This particulate cytoplasmic staining is consistent with the presence of GH

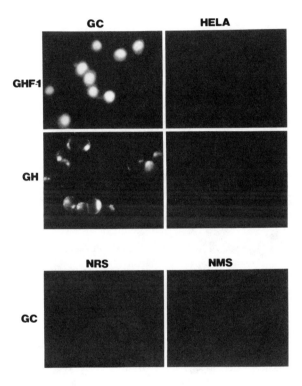

Figure 4: Nuclear localization of GHF-1 using a specific antibody

Indirect immunofluorescence microscopy using a specific peptide-generated antibody to the GHF-1 (anti-GHF-1) and a specific GH-antiserum (anti-GH). GC cells were also incubated with normal rabbit serum (NRS) and normal monkey serum (NMS) as negative controls for the specificity of anti-GHF-1 and anti-GH, respectively. Magnification 450 X.

in the secretory vesicles and the Golgi compartments within the cytoplasm. The nuclear staining of GC cells by the anti-GHF-1 antiserum appears very specific and was not observed if the cells were incubated with preimmune serum (NRS), or in other control experiments in which the anti-GHF-1 antiserum was blocked by preincubation with MB1 peptide originally used as immunogens (data not shown).

GHF-1 Contains a Homeobox

To identify an evolutionary conserved structural motif that may serve as the DNA-binding domain of GHF-1, we searched the Genetic Sequence data bank (PIR protein library version 16, March 1988). This analysis revealed a number of homeobox containing proteins, including EVE (Macdonald et al., 1986), IAB-7 (Regulski et al., 1985), MAT-al (Astell et al., 1981) and PRD (Frigerio et al., 1986), that share a considerable degree of sequence similarity (51-55%) with the C-terminal region of GHF-1. This homology was limited to the evolutionary conserved homeobox region of these proteins (Bodner et al. 1988). Like most other homeobox containing proteins, the homeobox of GHF-1 is near its C-terminus. The highest degree of sequence conservation is found at the C-terminal one-third of the GHF-1 homeobox. Comparative sequence analysis has indicated that this is also the most highly conserved region amongst all known homeoboxes (Gehring, 1987; Burglin, 1988). This region was proposed to represent one α-helix of a helix-turn-helix motif that may be involved in sequence recognition (Gehring, 1987). Overall, the GHF-1 homeobox exhibits 78% sequence identity to a consensus sequence derived by comparison of a large number of homeoboxes from different species (figure 5). If conservative amino acid changes are taken into consideration, the extent of homology reaches 85% (Figure 5). The GHF-1 homeobox deviates from the consensus sequence only at 4 out of 43 positions. Interestingly, the extent of similarity between GHF-1 and other mammalian homeobox containing proteins, whose functions are still unknown, is rather low (not shown).

## Discussion

GHF-1 is a transcription factor that binds to two sites within the GH promoter, which serve as essential promoter elements in vivo (Lefevre et al., 1987) and in vitro (Bodner and Karin, 1987). Previous studies have indicated that GHF-1 binding and transcriptional stimulatory activity are restricted to GH-expessing cells (Lefevre et al., 1987; McCormick et al. 1988). However, the basis for that specificity was unknown. GHF-1 transcripts can be detected only in pituitary RNA samples (Bodner et

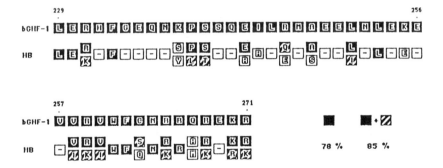

Figure 5: GHF-1 contains a homeobox

The sequence of the b-GHF-1 homeobox is shown in comparison to the consensus sequence compiled by analysis of a large number of homeobox-containing proteins. Solid black or hatched boxes indicate identical or similar amino acids, respectively. The percentage of sequence identity and similarity is shown.

al., 1988). When examined by immunofluorescence using a specific anti-GHF-1 antiserum, only the GH- and PRL-expressing cells of the pituitary appear to contain GHF-1 (Bodner et al., 1988). Therefore, the basis for the cell type specific expression of GH appears to be due to the very restricted distribution of the GH-specific transactivator, GHF-1.

Although the majority of the lactotropes do not express GH at one given time, various studies indicate that they are capable of transcribing that gene and therefore can be considered as committed to GH expression. Perhaps the strongest line of evidence is provided by the transgenic mice experiments. A chimeric gene consisting of the rGH promoter fused to hGH structural sequences introduced into the germ line of mice, was expressed in both somatotropes and lactotropes (Behringer et al., 1988). Other studies suggest that during the ontogeny of the anterior pituitary in rodents, the PRL expressing cells are derived from GH expressing cells (Slabaugh et al., 1982; Hoeffler et al., 1985). It has also been suggested that, under the influence of estrogen, the somatotropes can acquire the morphological characteristics of lactotropes (Strattmann et al., 1974), an effect that appears to be reversible and could account for the sex-linked differences in PRL and GH production (Leong et al., 1985). While these results are also consistent with sequential expansion and retraction of two separate clonal populations of cells, sequential hemolytic plaque assays of cultured rat neonatal pituitary cells indicated that almost all of the lactotropes that relesed PRL in one sequence were found to secrete GH in the other and therefore can be considered as mammosomatotropes. Only 1.3-2.5% of the PRL secreting cells were never found to secrete GH. In contrast, only a minority of somatotropes were ever observed to secrete PRL and the large majority of these cells only secreted GH (Hoeffler et al., 1985). In agreement with these observations, Behringer et al.(1988) found that expression of a

GH-DT chimeric gene in transgenic mice leads to ablation of GH-expressing cells as well as most of the PRL-expressing cells. These and the
previous morphological observations led to the development of a model
describing the lineal relationship among the progenitors of somatotropes
and lactotropes in rodents. According to this model, a stem cell
synthesizing neither GH nor PRL gives rise to cells that synthesize GH.
These cells give rise to mature somatotropes synthesizing only GH, which
can not undergo further differentiation, and to mammosomatotropes
expressing both GH and PRL. The latter cell type gives rise to mature
lactotropes that produce PRL. Rarely will the original stem cell give
rise to cells capable of synthesizing only PRL (Behringer et al., 1988).

Our results concerning the expression of GHF-1 are largely in agreement
with this model. We find that although GHF-1 does not bind to the PRL
promoter and probably is not directly required for the activation of
that gene (Bodner et al. 1988), it is present in both GH and PRL
expressing cells. These findings taken together with the known ability
of GHF-1 to activate transcription of the GH gene in vitro (Bodner and
Karin, 1987) suggest that GHF-1 is an essential factor for the differentiation of the somatotropic lineage. While so far, GH and PRL are the
only markers of this lineage that have been cloned (Miller and Eberhardt, 1983), other genes that should be specifically expressed in these
cell types are the genes that code for cell surface receptors for
GH-releasing factor (GRF), somatostatin and thyrotropin releasing
hormone (TRH) receptors. These hypothalamic hormones are known to
control the synthesis and release of GH and PRL (Seifert et al., 1985).
Thus, it will be important to isolate these genes and determine whether
they are coordinately under GHF-1 control.

If GHF-1 is indeed required for the differentiation of the somatotropic lineage in the pituitary, then its expression during embryogenesis should precede that of GH. The immunofluorescence localization approach employed here should also provide for detection of the first cells to express GHF-1 during the development of the anterior pituitary. While GHF-1 appears to be the only pituitary specific factor required for activation of the GH promoter (Bodner and Karin, 1987), it is possible that it acts in conjunction with other pituitary specific trans-activators operating on yet to be identified distal enhancer elements of this gene. In contrast to the trans-activation of the GH promoter region by GHF-1, no specific factors binding to this region, which are responsible for its repression in GH-nonproducing cells, were detected (McCormick et al. 1988). However, it is possible that a yet to be found negative factor may be responsible for a temporal shut-off of GH expression in lactotropes engaged in PRL expression.

The results listed above suggest that the mere appearance of GHF-1 in the developing cells of the anterior pituitary may be sufficient for activation of GH expression. According to this model, ectopic expression of GHF-1 in a different cell type should lead to activation of the GH gene and any other somatotrope specific gene controlled by GHF-1. This prediction can be tested by fusion of the GHF-1 cDNA to heterologous promoters of different specificity and introduction of such chimeric genes into tissue culture cells and the germ line of transgenic mice.

While the control of GH expression by GHF-1 appears to be quite clear, the present studies raise the important recurring question: What controls the expression of GHF-1? We found that extinction of GH expression in somatic cell hybrids is mediated by the supression of GHF-1 expression (McCormick et al. 1988). Regardless of the mechanism responsible for repression of GHF-1 expression, it is clear from these studies that fibroblasts and probably other GH-nonexpressing cell types express an extinguisher responsible for the repression of GHF-1.    In addition, we can expect to find at least one positive factor (PRLF) responsible for the activation of prolactin gen and repression of growth-hormone gene at a critical point during the development of the pituitary (see "Two factors-Two cell types" model in figure 6).    We hope that by studying the regulation of GHF-1 expression, it will be

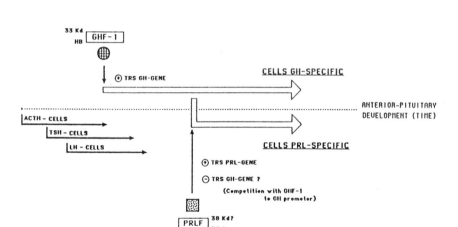

Figure 6: "Two factors-two cell types" pituitary development model

The aparition of the transcription factor GHF-1 during the anterior-pituitary development activate the specific expression of the GH gen. Later in the time, the aparition of a second new transcription factor (PRLF ?) activate the expression of the prolactin (PRL) gen and repress the GH transcription likely by affinity-competition with GHF-1 to the binding sites of the GH promotor.

possible to understand the switch from GH-cells to PRL-cells in the anterior pituitary development. Such an approach should lead to elucidation of the regulatory network controlling the development of the anterior pituitary.

The remarkable specificity of GHF-1 expression is quite different from the expression patterns of most other transcription factors which appear to be present ubiquitously in most cell types (see Jones et al., 1988 for review). However, it is very similar to the expression of most of the homeobox containing genes in Drosphila to which GHF-1 is related. For example, the en, ftz and Ubx gene products were shown to accumulate in the same cell layers and segments whose formation is affected by mutations in these genes (Dinardo et al., 1985; Caroll and Scott, 1985; Beachy et al., 1985). Thus, like GHF-1, their site of expression appears to be identical to their site of action. It is also known that regulatory interactions between homeobox containing genes are respon-sible for limiting their expression to specific sites (see Gehring, 1987; Scott and Caroll, 1987 for reviews). Thus, it will be of interest to determine whether the genes that control GHF-1 expression will also be members of the homeobox family. Another interesting property exhibited by homeobox containing genes in Drosophila is that their expression sites in adult flies are different from the sites in which they are expressed in embryos (Gehring, 1987; Scott and Caroll, 1987). While mutational analysis indicates that these genes play very important roles in embryogenesis, their roles in the formation of adult structures are not all that clear. Because of its homeobox homology, it will be important to compare the embryonic pattern of GHF-1 expression to its expression pattern in adult animals described here. Since other homeobox containing proteins like EVE are capable of recognizing two different sets of binding sites (Hoey and Levine, 1988) and others like

MATa1 can interact with other proteins (MATα2) to form heterodimers with
new sequence specificity (Goutte and Johnson, 1988), it is conceivable
that a protein like GHF-1 may play more than a single role during
development.


## Acknowledgments

We are indebted to Paul Matsudaira (Whitehead Institute) for performing
the amino acid sequence determination of GHF-1 peptides, , Graeme Bell
(University of Chicago) for the bovine pituitary cDNA library, Rob
Richards for the synthetic peptides, Yagha Sinha  for GH antibodies,
Thomas Deerinck and Mark Ellisman for helping with the IF microscopy,
David Wu for skillful technical assistance, Donna Caruso for dedicated
preparation of this manuscript and Victoria Vila for critical reading
and helpful discussions.  J. L. C. and M. B. were supported by EMBO long
term fellowships and L. T. by a Danish Medical Research Council fellow-
ship.  Work was supported by Public Health Service Grants DK-38527.

## References

Astell, C. R., Ahlstrom-Jonasson, L., Smith, M., Tatchell, K., Nasmyth,
    K. A. and Hall, B. D. (1981) The sequence of the DNAs coding for the
    mating-type loci of Saccharomyces cerevisiae. Cell 27, 15-23
Bancroft, F. C. (1981) GH cells: functional clonal lines of rat pituit-
    ary tumor cells. In: Functionally differentiated cell lines, (ed. G.
    Sato), pp 47-55. Alan R. Liss, New York.
Beachy, P. A., Helfand, S. L. and Hogness, D. S. (1985) Segmental
    distribution of bithorax complex proteins during Drosophila
    development. Nature 313, 545-551.
Behringer, R. R., Mathews, L. S., Palmiter, R. D. and Brinster, R. L.
    (1988) Dwarf mice produced by genetic ablation of growth hormone-
    expressing cells. Genes and Development 2, 453-460.
Bodner, M. and Karin, M. (1987) A pituitary-specific trans-acting
    factor can stimulate transcription from the growth hormone promoter
    in extracts of non-expressing cells. Cell 50, 267-275.

Bodner, M., Castrillo, J. L., Theill, L. E., Deerinck, T., Allisman, M. and Karin,M. (1988) The pituitary specific transcription factor GHF-1 is a homeobox containing protein. Cell in press.

Bodner, M., Castrillo, J. L. and Karin, M. (1988) Purification and characterization of the pituitary specific transcription factor, GHF-1. Science submitted.

Burglin, T. R. (1988) The yeast regulatory gene PHO2 encodes a homeo box. Cell 53, 339-340.

Caroll, S. B. and Scott, M. P. (1985) Localization of the fushi tarazu protein during Drosophila embryogenesis. Cell 43, 47-57.

Chatelain, A., Dupouy, J.P. and Dubois, M.P. (1979). Ontogenesis of cells producing polypeptide hormones (ACTH, MSH, LPH, GH, Prolactin) in the fetal hypophysis of the rat: Influence of the hypothalamus. Cell Tissue Res. 196, 409-427.

DiNardo, S., Kuner, J. M., Theis, J. and O'Farrell, P. H. (1985) Development of embryonic pattern in D.melanogaster as revealed by accumulation of the nuclear engrailed protein. Cell 43, 59-69.

Engvall, E. (1980) Enzyme immunoassay ELISA and EMIT. Methods Enzymol 70, 419-439.

Frigerio, G., Burri, M., Bopp, D., Baumgartner, K. and Noll, M. (1986) Structure of the segmentation gene paired and the Drosophila PRD gene set as part of a gene network. Cell 47, 735-746.

Gehring, W. J. (1987) Homeoboxes in the study of development. Science 236, 1245-1252.

Goutte, C. and Johnson, A. D. (1988) a1 protein alters the DNA binding specificity of αα repressor. Cell 52, 875-882.

Hattori, M. and Sakaki, Y. (1986) Dideoxy sequencing method using denatured plasmid templates. Anal. Biochem. 152, 232-238.

Hoeffler, J.P., Boockfor, R.R. and Frawley, S. (1985). Ontogeny of prolactin cells in neonatal rats: Initial prolactin secretors also release growth hormone. Endocrinology 117, 187-195.

Hoey, T. and Levine, M. (1988) Divergent homeo box proteins recognize similar DNA sequences in Drosophila. Nature 332, 858-861.

Ivarie, R. D., Schaeter, B. S. and O'Farrell, P. H. (1983) The level of expression of the rat growth-hormone gene in liver tumor cells is at least eight orders of magnitude less than that in anterior pituitary cells. Mol. Cell. Biol. 3, 1460-1467.

Jones, N. C., Rigby, P. W. J. and Ziff, E. B. (1988) Trans-acting protein factors and the regulation of eucaryotic transcription: lessons from studies on DNA tumor viruses. Genes and Development 2, 267-281.

Lefevre, C., Imagawa, M., Dana, S., Grindlay, J., Bodner, M. and Karin, M. (1987) Tissue-specific expression of the human growth hormone gene is conferred in part by binding of a specific trans-acting factor. EMBO J. 6, 971-981.

Leong, D.A., Lau, S.K., Sinha, Y.N., Kaiser, D.L. and Thorner, M.O. (1985). Enumeration of lactotropes and somatotropes among male and female pituitary cells in culture: Evidence in favor of a mammosomatotrope subpopultion in the rat. Endocrinology 116, 1371-1378.

Macdonald, D. M., Ingham, P. and Struhl, G. (1986) Isolation, structure and expression of even-skipped: A second pair-rule gene of Drosophila containing a homeo box. Cell 47, 721-734.

Maniatis, T., Fritisch, F. F. and Sambrook, J. (1982) Molecular cloning, a laboratory manual. Cold Spring Harbor, New York.

Matsudaira, P. (1987) Sequence from picomole quantities of proteins electroblotted onto polyvinylidene difluoride membranes. J. Biol. Chem. 262, 10035-10038.

McCormick, A., Wu, D., Castrillo, J. L., Dana, S., Strobl, J., Thompson, E. B. and Karin, M. (1988). Extinction of growth hormone expression in somatic cell hybrids involves supression of the specific trans-activator, GHF-1. Cell in press.

Miller, W. L. and Eberhardt, N. L. (1983) Structure and evolution of the growth hormone gene family. Endocrine Rev. 4, 97-130.

Mosbach, K., Guilford, H., Ohlsson, R. and Scott, M. (1972) General ligands in affinity chromatography. Biochem. J. 127, 625-631.

Regulski, M., Harding, K., Kostriken, R., Karch, F., Levine, M. and McGinnis, W. (1985) Homeo box genes of the antennapedia and bithorax complexes of Drosophila. Cell 43, 71-80.

Scott, M. P. and Caroll, J. B. (1987) The segmentation and homeotic gene network in early Drosophila development. Cell 51, 689-698.

Seifert, H., Perrin, M., Rivier, J. and Vale, W. (1985) Binding sites of growth hormone releasing factor on rat anterior pituitary cells. Nature 313, 487-489.

Slabaugh, M. B., Lieberman, M. E., Rutledge, J. J. and Gorski, J. (1982) Ontogeny of growth hormone and prolactin gene expression in mice. Endocrinology 110, 1489-1497.

Strattmann, I. E., Ezrin, C., Sellers, E. A. (1974) Estrogen-induced transformation of somatotrophs into mammotrophs in the rat. Cell Tissue Res. 152, 229-238.

Tashjian, A. H., Yasumura, Y., Levine, L., Sato, G. H. and Parker, M. L. (1968) Establishment of clonal strains of rat pituitary tumor cells that secrete growth hormone. Endocrinology 82, 342-352.

DISCUSSION OF THE PAPER PRESENTED BY J.L. CASTRILLO

JUMP: What is the copy number of the GHF-1 transcription factor per cell?

CASTRILLO: Actually, we don't know the copy number of GHF-1. We are now trying to compare this number in different cell lines.

JUMP: Do you anticipate GHF interaction with genes other than the growth hormone gene in GC cells?

CASTRILLO: No, the binding of GHF-1 is highly specific.

JUMP: The amount of immunostaining for GHF-1 of pituitary cells was considerably less than the GC cells. Please comment.

CASTRILLO: GC (or GH3) cells are transformed cells and must express more GHF-1 than pituitary cells. That helped us to purify and clone our factor.

THOMPSON: Please discuss the controversy regarding Pit 1 and GHF-1 binding to the GH and Prolactin gene promoters. Are there two or only one factors?

CASTRILLO: Our affinity preparations of GHF-1 and the trpE fusion proteins of the cloned factor bind only to the GH promoter. You have to realize that it is not the Pit-1 of Rosenfeld's group. The prolactin factor(s) same as binding sites are A/T rich, and if we add 10-100 times more protein maybe it will be possible to bind non-specifically to these sites. Thus, it is very important to know the ratio DNA/protein in all of these experiments. In addition, the same trouble happened with the trans-activation in CAT-assays. If we add high levels of the factor it is possible to trans-activate any promoter.

CIDLOWSKI: What happens to GH expression if you transfect GF1 into cells which normally do not produce GH?

CASTRILLO: If we made the experiments with exogenous GH-CAT plasmids, it is possible to detect trans-activation of the GH promoter. We have to perform the experiments to detect the endogenous expression of the GH gene.

TSAI: Since your GHF-1 DNA complex cannot be completely shifted by antibody, is it possible GHF-1 is heterogenous?

CASTRILLO: Our purified GHF-1 is a doublet of 33-32 Kd, antigenically related. At this moment, we don't know what kind of modification is involved.

ROY: Have you looked for localization of the transacting factors for GH gene in the TSH producing cells? The reason I ask this is that a small percent of the thyrotrops also produce GH.

CASTRILLO: We could not because we do not have a specific antibody anti-rat TSH.

DISCUSSANTS: D. JUMP, J.L. CASTRILLO, E.B. THOMPSON, J. CIDLOWSKI, M.J. TSAI AND A.K. ROY.

# Regulation of EGF Receptors and Nuclear Protooncogenes by Estrogen

D.S. Loose-Mitchell[*], C. Chiappetta[*], R.M. Gardner[*],

J.L. Kirkland[+], T.-H. Lin[+], R.B. Lingham[*],

V.R. Mukku[*], C. Orengo[*], and G.M. Stancel[*]

[*]Department of Pharmacology, The University of Texas Medical School at Houston, Houston, Texas  77225 and [+]Division of Endocrinology, Department of Pediatrics, Baylor College of Medicine, Houston, Texas  77030

## Introduction

Estrogens stimulate the growth of both normal target tissues and estrogen sensitive tumor cells.  Recent studies from a number of laboratories have raised the possibility that these growth responses may involve an interplay between the steroid, peptide growth factors, and nuclear protooncogenes.  For example, estrogens stimulate secretion of IGF and EGF-like peptides in MCF-7 human breast cancer cells, and these peptides can support MCF-7 cell growth in the absence of the steroid (Dickson et al., 1986a, b).  Similarly, in vivo, estrogens regulate uterine EGF (Gonzales et al.; DiAugustine et al.), EGF receptors (Mukku et al., 1985a), IGF (Murphy et al., 1988; Murphy et al., 1987a) and several nuclear protooncogenes (Travers et al.;

Murphy et al., 1987b; Loose-Mitchell et al.; Weisz et al.).
Furthermore, it has been reported that antibodies to EGF
can diminish estrogen induced uterine growth in organ
cultures   (McLachlan et al.).  Consequently, our laboratories have studied the in vivo induction of the nuclear
protooncogene c-fos (Loose-Mitchell et al.) and the EGF
receptor (Lingham et al.; Mukku et al., 1985a) by estrogens.    c-fos is the cellular analog of the v-fos oncogene
originally identified in osteogenic sarcoma viruses.  Early
studies in a variety of systems demonstrated that fos expression is rapidly and dramatically increased by serum,
peptide growth factors, and other agents which promote
growth such as tumor promoting phorbol esters (Alt et al.).
Recent "antisense" studies have shown an essential role for
fos in the mitogenic response of 3T3 cells (Holt et al.;
Nishikura et al.).  While the precise role of fos in the
growth response is not clear, recent studies have suggested
that c-fos is a trans-acting transcriptional regulator of
gene expression (Ruther et al.; Lech et al.; Distel et al.;
Rauscher et al.,; Chiu et al.).

Despite evidence for a role of fos in cell growth, it is
clear that fos expression alone is not sufficient to stimulate a mitogenic response.  For example, growth factors
such as PDGF, which stimulate fos expression, are not
capable of stimulating DNA synthesis without the coordinate
action of other growth factors such as EGF or IFG-1 (Stiles
et al.).  As a consequence, we have examined the effect of
estrogen administration on uterine EGF receptor levels and
others have studies effects of the hormone on this growth

factor itself.

EGF was originally isolated by Cohen as a peptide capable of stimulating incisor eruption and eyelid opening in newborn mice (Cohen). Subsequent studies showed that the EGF receptor is a plasma membrane protein with a tyrosine kinase activity which is stimulated by ligand binding (Carpenter, 1987). More recently, it has been shown that the EGF receptor is the cellular analog of the v-erb B oncogene of avian erythroblastosis virus (Downward et al.). It is now recognized that EGF can stimulate the growth of many cell types both in vivo and in vitro (Carpenter et al., 1979). It thus seemed reasonable to investigate the possibility that EGF receptor regulation might be involved in the stimulation of target tissue growth by estrogens.

## Results

### Regulation of c-fos Expression by Estrogen

As shown in Figure 1, estradiol administration to immature female rats causes a rapid and dramatic increase in c-fos expression measured by RNA blot analysis. The level of c-fos mRNA doubles within 30 minutes and reaches a maximum level in 3 hours. While there is some variation in the exact increases seen in different experiments, we routinely observe increases of 25-50 fold. Other studies have shown that this induction of fos is: (1) specific for estrogenic steroids; (2) sensitive to actinomycin D; and (3) insensitive to puromycin. Also, induction occurs within the same dose range of the steroid that stimulates uterine DNA synthesis (Loose-Mitchell et al.).

In addition to c-fos, it has also been shown by others that

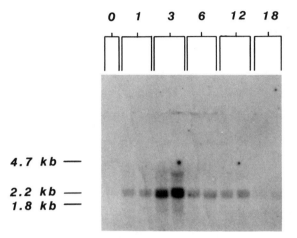

Figure 1. Induction of c-fos expression. Total uterine
RNA was isolated at the indicated times (in hours) after
estradiol treatment, and used for Northern analysis of the
2.2 kb c-fos transcript. Data taken from Loose-Mitchell,
et al., with permission.

estradiol administration stimulates expression of the
nuclear protooncogene c-myc (Travers et al.), and we have
confirmed this observation. Thus, estrogens appear to
rapidly stimulate expression of several nuclear protooncogenes as part of the normal in vivo growth response of
the uterus.

The sharp increase and the equally sharp decline in c-fos
mRNA levels suggested to us that both the level of c-fos
and the timing of its expression might be precisely
controlled during the tissue growth response. To test this
possibility we investigated whether c-fos expression could

undergo multiple "rounds" of stimulation by estradiol.
In this series of experiments animals were treated initially with a dose of 4 µg/kg of estradiol for various times
(e.g., 6, 12, 18 hours) past the peak of c-fos mRNA accumulation (see Figure 1). This dose was chosen because it oc-
cupies less than half of the available estrogen receptor
sites in uterine tissue (Anderson et al., 1972, 1973).
After the level of c-fos mRNA had peaked and fallen, the
animals were injected with a second dose of estradiol and
sacrificed three hours later. Northern analyses of RNA
prepared from these tissues indicated that the second dose
of the hormone did not appreciably increase c-fos or c-myc
mRNA levels, i.e., the expression of these two gene products became refractory to repetitive hormonal stimulation
(Orengo, et al., unpublished observations). This finding
suggests that a cellular mechanism exists to prevent a
secondary increase in nuclear protooncogene expression in
response to estrogen. At present we do not have any evidence indicating the level at which this mechanism operates, but we are actively pursing this question.
It is interesting to note, however, that c-fos and c-myc
induction by estrogen occurs normally if the second injection of estradiol is spaced more than 24 hours after the
first (Orengo, et al., unpublished observation). This may
be significant because uterine DNA synthesis peaks 24 hours
after estrogen treatment (Mukku et al., 1982).

Properties of Uterine EGF Receptors

As illustrated in Figure 2, membranes prepared from the
uterus of immature rats contain a single class of saturable

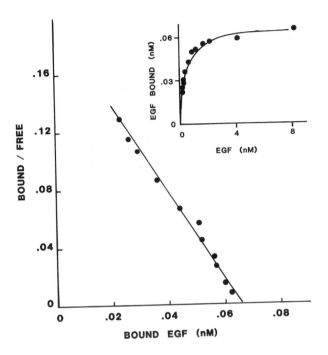

Figure 2. EGF receptors in uterine membranes. The specific binding of $^{125}$I-EGF to uterine membranes was measured
by a filter binding assay. Taken from Mukku, et al.,
1985b, with permission.

binding sites for EGF. In a number of experiments we
(Mukku et al., 1985a,b) have measured a $K_d$ value of 0.1 -
1 nM for these sites. These binding sites are present in
all major uterine cell types in the rat (Lin et al.) and in
the human (Chegini et al.). Other studies revealed that the
MW of the uterine EGF receptor is 170,000, and that this
receptor contains a tyrosine kinase activity which can be
observed by autophosporylation and the use of exogenous

substrates (Mukku et al., 1985a,b). The uterine EGF receptor thus appears similar to receptors in a variety of other
normal tissues and tumor cells (Carpenter, 1987).

Regulation of EGF Receptor Levels by Estrogen

Treatment of immature animals with estradiol leads to an
increase in uterine levels of EGF receptor mRNA as illustrated in Figure 3. EGF receptor mRNA levels rise between
1 and 3 hours after hormone administration, remain at peak
levels (4-5 fold above controls) for another 3 hours, and
then decline between 6 and 12 hours after treatment. This
induction is specific for estrogenic steroids, is sensitive
to actinomycin D but not puromycin, and is dose dependent
(Lingham et al.).

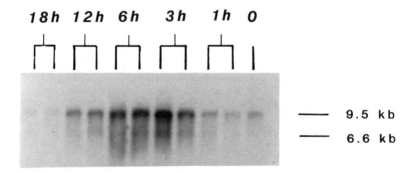

Figure 3. Induction of EGF receptor mRNA. Total uterine
RNA was isolated at the indicated times after estradiol
treatment and used for Northern analysis of the 9.5 and 6.6
kb EGF receptor transcripts. Taken from Lingham, et al.,
with permission.

At slightly longer times after estrogen administration (6-12 hours) there is a 3-fold increase in functional EGF receptors measured by ligand binding or EGF stimulated tyrosine kinase activity. Inhibitor studies have suggested that this increase in receptor levels represents de novo protein synthesis resulting, at least in part, from transcriptional activation (Mukku et al., 1985a). The time courses for the induction of EGF receptor mRNA and functional EGF receptors are shown in Figure 4.

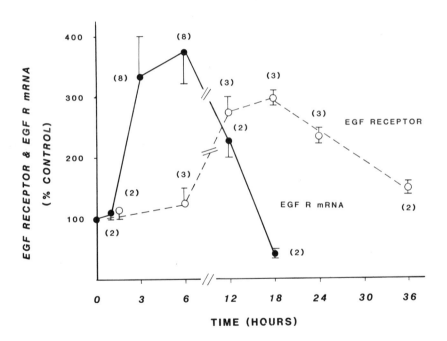

Figure 4. Induction of EGF receptor and its mRNA. Uterine EGF receptor mRNA (solid line) and functional EGF receptor (dashed line) levels after estradiol treatment. Taken from Lingham, et al., with permission.

The studies above were all performed using immature rats to monitor the effects of estrogen on uterine EGF receptor levels.  To insure that these effects were not limited to this animal model, we performed similar studies in the immature mouse (Gardner, et al., in preparation) and the mature, castrated rat (Gardner, et al., 1989).  In both systems estradiol produces increases in functional EGF receptors similar to that seen in the immature rat.  Also, we have shown recently that in the mature cycling rat, EGF receptor levels vary throughout the estrous cycle in parallel with plasma estrogen levels and uterine levels of occupied nuclear estrogen receptors (Gardner, et al., 1989).  Thus, it appears that estrogens regulate EGF receptor levels in several species and at several developmental stages.

Regulation of Uterine Contractility by EGF

Our initial interest in EGF and its receptor arose because of the possibility that this peptide might play a role in estrogen stimulated uterine growth.  A review of the literature, however, revealed that EGF has other biological activities besides the stimulation of cell growth.  For example, EGF inhibits gastric acid secretion (Gregory) and stimulates contractions of vascular smooth muscle (Muramatsu et al.; Berk et al.).  This latter report and the observation that the myometrium (as well as other uterine cell types) contains EGF receptors (Lin et al.; Chegini et al.), prompted us to investigate the possibility that EGF might affect uterine contractility.

For these studies, segments of uterine tissue from estrogen primed rats were suspended in vitro and the contractile response was measured following EGF addition. As seen in Figure 5, addition of EGF caused a prompt response. Within minutes after EGF addition there is an increase in both tone and rhythmicity, and this effect is sustained for several hours (Gardner et al., 1987).

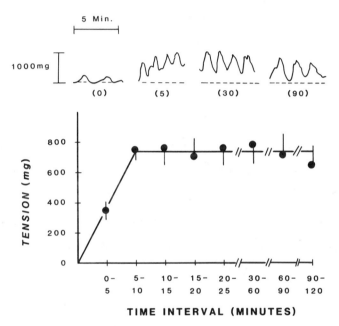

Figure 5. EGF induction of uterine contractions. Segments of uterine tissue were suspended in vitro and the tension which developed following EGF addition was measured. Top - tracing from an individual sample at 0, 5, 30 and 90 minutes after EGF addition. Bottom - maximum tension developed in response to EGF at the indicated times for a series of samples (N=7/point). Taken from Gardner, et al., 1987, with permission.

Other experiments revealed that this effect: (1) requires estrogen priming in vivo; (2) occurs in tissues from mature and immature animals; (3) occurs at EGF concentrations ($ED_{50}$ = 3.5 nM) consistent with a receptor mediated effect; and (4) is abolished by anti-EGF antibodies (Gardner, et al., 1987). More recent studies suggest that this effect of EGF is mediated by a mobilization of arachidonic acid and production of uterotonic metabolites by both cyclooxygenase and lipoxygenase catalyzed pathways (Gardner et. al., 1988).

## Discussion

### Induction of c-fos Expression by Estrogen

Estrogen treatment in vivo causes a rapid and dramatic increase in c-fos mRNA levels. It seems most likely that this effect results, at least in part, from direct activation of fos transcription by the estrogen receptor, since: (1) the induction is rapid, specific,and sensitive to inhibitors of transcription but not translation (Loose-Mitchell et al., 1988); (2) a palindromic sequence (5' - GGTCT ... AGACC - 3') is located at -219/-207 from the start site of fos transcription (Treisman) and this sequence is very similar to the sequence 5' - GGTCA ... TGACC - 3' of the estrogen responsive element of the Xenopus vitellogenin $A_2$ gene (Klein-Hitpass et al.); and (3) estrogens stimulate an increase in fos expression measured directly by nuclear run-on assays (Weisz et al.). Fos may thus be a excellent marker for very early genomic responses to estrogen in normal uterine tissue.

While the increase in c-fos after estrogen treatment is

striking, the role of fos in uterine growth in less clear.
Among several possibilities, two actions of fos on uterine
growth seem most likely. First, fos could function as part
of a cascade to amplify nuclear responses resulting from
the initial interaction of the estrogen receptor complex
with estrogen responsive elements in the genome. For
example, fos interacts with the nuclear transcription
factor AP-1 to stimulate transcription of other genes
(Chiu et al.). Therefore, a direct increase in fos by the
estrogen receptor could initiate a cascade which would
alter the expression of genes that do not contain estrogen
responsive elements. Such "recruitment" of genes not
directly regulated by the estrogen receptor would seem
ideally suited to support the massive growth of the uterus
in response to the steroid. Furthermore, such a cascade
would have a temporal aspect, i.e., the rapid increase and
decline in fos expression would effect the timing of specific genomic responses during a defined time span of the
growth response. In this regard, fos could be a "third
messenger" for steroid induced responses, similar to the
"third messenger" function of fos proposed by Verma and
Sassone-Corsi (1987) for transmission of signals from cytoplasmic second messengers to the nuclear machinery of the
cell.

A second intriguing possibility is that fos expression, as
well as translational and post-translational regulation of
fos, could serve to integrate and/or modulate signals
arising from the estrogen-receptor complex with signals
from receptor systems for other regulatory molecules. This

could conceivably happen directly at the level of fos regulation, since many agents (e.g., serum, specific growth factors, cyclic nucleotides, phorbol esters, calcium, etc.) can regulate fos at transcriptional, translational and post-translational levels (Alt et al.). Another possibility is an interaction at the level of those genes regulated by fos. For example, the regulation of gene expression by the AP-1/fos system may be via the DNA sequence which confers phorbol ester/calcium sensitivity to specific genes (Ruther et al.; Lech et al.; Distel et al.; Rauscher et al.; Chiu et al.). This would allow for "downstream" communication between estrogen receptor and other signalling mechanisms.

Regulation of EGF Receptors by Estrogen

The uterus contains both EGF receptors (Mukku et al., 1985a,b) and EGF (Gonzales et al.; DiAugustine et al.), and EGF is also found in the uterine luminal fluid (Imai). More importantly, EGF stimulates the growth of uterine epithelial cells in cell culture (Tomooka et al.) and in organ culture (McLachlan et al.). These results suggest strongly that EGF may be an important regulator of uterine growth in vivo. Our studies have shown that estrogen regulates the level of uterine EGF receptors. This effect may be at the transcriptional level since the increase in functional receptors is preceded by an increase in the receptor mRNA, and this effect is blocked by transcriptional, but not translational, inhibitors. The time course, hormonal specificity, and dose response profiles for this effect indicate a correlation between the increase

in functional EGF receptors and estrogen stimulated DNA
synthesis (Mukku et al., 1985a; Lingham et al.).
If, as seems likely, EGF plays a role in uterine growth,
regulation of its cognate receptor could play a role in the
growth response by one of two general mechanisms. First,
an increase in EGF receptor levels, alone or in combination
with an increase in EGF, might be necessary to generate a
threshold level of a second messenger(s) required for cell
growth. Second, growing cells may require increased synthesis of growth factor receptors to insure that down
regulation does not deplete receptor levels below some
critical level. This would be consistent with the observations that: (1) EGF stimulated growth of fibroblasts correlates with the steady-state, rather than the initial,
levels of EGF-receptor complexes present (Knauer et al.),
and (2) fibroblasts must be exposed to EGF for extended
periods of time to obtain a maximum mitogenic response
(Carpenter et al., 1976). These two possibilities are not
mutually exclusive.

As Overall Model for Estrogen Stimulated Uterine Growth

It has been clear for some time that the overall control of
uterine growth by estrogens involves a series of regulatory
steps rather than an initial, transient interaction of this
hormone with its receptor (Anderson et al., 1972, 1973).
In a general sense, the same concept of multiple régulatory
steps applies to fibroblast growth (Stiles et al.). In the
cultured fibroblast systems, a "competence" factor (e.g.,
PDGF) is required to initiate growth by allowing quiescent
cells in $G_0$ to enter $G_1$. Competence factors such as PDGF

appear to act by rapid stimulation of myc, fos and other so-called "competence" (Stiles et al.) or "immediate early" genes (Lau et al.). However, competence factors alone are incapable of stimulating mitogenesis - fibroblasts require one or more other factors to progress through $G_1$ and enter S phase. Examples of "progression factors" include EGF and IGF-1. When viewed in this way, there is an apparent similarity to estrogen mediated uterine growth. First, estrogen stimulates c-fos (Weisz et al.; Loose-Mitchell et al.) and c-myc (Travers et al.; Murphy et al., 1987a; Weisz et al.), and may well stimulate other competence genes. Second, estrogen stimulates IGF-1 mRNA levels (Murphy et al., 1987a) as well as EGF (McLachlan et al.) and EGF receptors (Mukku et al., 1985a; Lingham et al.) in the uterus.

We would thus propose the following working hypothesis for estrogen mediated uterine growth. In the initial phase of estrogen action, the steroid acts via its receptor to rapidly stimulate expression of c-fos and other genes containing estrogen responsive elements. Fos (and possibly other early products) amplifies this initial signal by transcriptional activation of other genes with distinct regulatory elements (e.g., the AP-1 dependant enhancer element). In concert with these effects, estrogen (directly and/or indirectly) stimulates the production of peptide growth factors and/or their cognate receptors. These factors then support continued uterine growth via autocrine or paracrine mechanisms. This working model is not meant to be exclusive, since many other regulatory signals (either extra or intracellular) could play important roles in the overall tissue response. Rather, our primary intention is to propose a testable model of normal growth in vivo, based upon our current understanding of estrogen action and growth control. Much work clearly remains to be done to prove or disapprove this model of estrogen stimulated uterine growth.

## Acknowledgements

We thank Ms. Martha Zurita for preparation of this manuscript. Work in our laboratories were supported by NIH grants HD-08615 (G.M.S.), DK-38965 (D.S.L.-M.) and RR-01685 (Bionet) for computer use, and a grant from the John P. McGovern Foundation (J.L.K.).

## Abbreviations

EGF-epidermal growth factor; IGF-1-insulin like growth factor; PDGF-platelet derived growth factor.

## References

Alt, F.W., Harlow, E., and Ziff, E.B. 1987. Nuclear oncogenes. Cold Spring Harbor Laboratory, Cold Spring Harbor, N.Y.

Anderson, J.N., Clark, J.H. and Peck, E.J., Jr. 1972. The relationship between nuclear receptor estrogen binding and uterotrophic responses. Biochem. Biophys. Res. Commun. 48:1460-1468.

Anderson, J.N., Peck, E.J., Jr. and Clark, J.H. 1973. Nuclear receptor estrogen complex: relationship between concentration and early uterotrophic responses. Endocrinology. 92:1488-1495.

Berk, B.C., Brock, T.B., Webb, R.C., Taubman, M.B., Atkinson, W.J., Ginbrone, M.A., and Alexander, R.W. 1985. EGF, a vascular smooth muscle mitogen, induces rat aortic contractions, J. Clin. Invest. 75:1083-1086.

Carpenter, G. 1987. Receptors for epidermal growth factor and other polypeptide mitogens. Ann. Rev. Biochem. 56:881-914.

Carpenter, G. and Cohen, S. 1976. [125]I-labeled epidermal growth factor: binding, internalization and degradation in human fibroblasts. J. Cell Biol. 71:159-171.

Carpenter, G., Cohen, S. 1979. Epidermal growth factor. Annu. Rev. Biochem. 48:193-216.

Chegini, N., Rao, C.V., Wakin, N., Sanfilippo, J. 1986. Binding of [125]I-epidermal growth factor in human uterus. Cell Tissue Res. 246:543-548.

Chiu, R., Boyle, W.J., Meek, J., Smeol, T., Hunter, T., and Karin, M. 1988. The c-fos protein interacts with c-Jun/AP-1 to stimulate transcription of AP-1 responsive genes. Cell 54:541-552.

Cohen, S. 1962. Isolation of a mouse submaxillary protein accelerating incisor eruption and eyelid opening in the newborn animal. J. Biol. Chem. 237:1555-1562.

DiAugustine, R.P., Lammon, D.E., McLachlan, J.A. 1985. Sex steroid hormones rapidly increase uterine epidermal growth factor (EGF). Endocrinology. 116:269 (abstr.).

Dickson, R.B., McManaway, M.E., Lippman, M.E., Lippman, M.E. 1986a. Estrogen-induced factors of breast cancer cells partially replace estrogen to promote tumor growth. Science 232:1540-1543.

Dickson, R.B., Huff, K.K., Spencer, E.M., and Lippman, M.E. 1986b. Induction of epidermal growth factor-related poly-peptides by 17 β -estradiol in MCF-7 human breast cancer cells. Endocrinology 118:138-142.

Distel, R.J., Ro, H.S., Rosen, B.S., Groves, D.L. and Spiegelman, B.M. 1987. Nucleoprotein complexes that regu-late gene expression in adipocyte differentiation: Direct participation of c-fos. Cell 49:835-844.

Downward, J., Yarden, Y., Mayes, E., Scrace, G., Totty, N., Stockwell, P., Ullrich, A., Schlessinger, J., Waterfield, M.D. 1984. Close similarity of epidermal growth factor receptor and v-erb-B oncogene protein sequences. Nature (London) 307:521-527.

Gardner, R.M., Lingham, R.B., and Stancel, G.M. 1987. Epidermal growth factor stimulates contractions of the isolated uterus. FASEB J. 1:224-228.

Gardner, R.M., Goldsmith, J.R., and Stancel, G.M. 1988. Pharmacological characterization of epidermal growth factor (EGF) induced uterine contractions. FASEB J. 2:A1143.

Gardner, R.M., Verner, G., Kirkland, J.L., and Stancel, G.M. 1989. Regulation of EGF receptors by estrogen in the mature rat and during the estrous cycle. J. Steroid Biochem. (in press).

Gonzalez, F., Lakshmanan, J., Hoath, S., Fisher, D.A.
1984. Effect of oestradiol-17β on uterine epidermal growth
factor concentration in immature mice. Acta Endocrinol.
(Copenhagen) 105:425-428.

Gregory, H. In vivo aspects of urogastrone-epidermal
growth factor. 1985. J. Cell Sci. Suppl. 3:11-17.

Holt, J.T., Gopal, T.V., Moulton, A.D., and Nienhuis, A.W.
1986. Inducible production of c-fos antisense RNA inhibits
3T3 cell proliferation. Proc. Nat. Acad. Sci. 83:4794-4798.

Imai, Y. Epidermal growth factor in rat uterine luminal
fluid. 1982. Endocrinology 110(Suppl): 162:(abstr.).

Klein-Hitpass, L., Schrorp, M., Wagner, U., Ryfell, G.U.
1986. An estrogen responsive element derived from the 5'
flanking region of xenopus vitellogenin A2 gene functions
in transfected human cells. Cell 46:1053-1061.

Knauer, D.J., Wiley, H.S., and Cunningham, D.B. 1984.
Relationship between epidermal growth factor receptor
occupancy and mitogenic response. Quantitative analysis
using a steady state model system. J. Biol. Chem.
259:5623-5631.

Lau, L.F. and Nathans, D. 1987. Expression of a set of
growth-related immediate early genes in BALB/c 3T3 cells:
coordinate regulation of c-fos or c-myc. EMBO J. 4:3145-
3152.

Lech, K., Anderson, K. and Brent, R. 1988. DNA-bound c-fos
proteins activate transcription in yeast. Cell 52:179-184.

Lin, T.S., Kirkland, J.L., Mukku, V.R., and Stancel, G.M.
1988. Autoradiographic localization of epidermal growth
factor binding in individual uterine cell types. Biol.
Reproduction 38:403-411.

Lingham, R.B., Stancel, G.M. and Loose-Mitchell, D.S. 1988.
Regulation of epidermal growth factor receptor messenger
RNA by estrogen. Mol. Endocrinol. 2:230-235.

Loose-Mitchell, D.S., Chiappetta, C., and Stancel, G.M.
1988. Estrogen regulation c-fos messenger ribonucleic
acid. Mol. Endocrinol. 2:946-951.

McLachlan, J.A., DiAugustine, R.P., and Newbold R.R. 1987.
Estrogen induced uterine cell proliferation in organ cul-
ture is inhibited by antibodies to epidermal growth factor.
Program of the 69th Annual Meeting of the Endocrine
Society, Indianapolis, IN, p. 99 (Abstract #13).

Mukku, V.R., Kirkland, J.L., Hardy, M., and Stancel, G.M.
1982. Hormonal control of uterine growth:  Temporal relationships between estrogen administration and DNA synthesis. Endocrinology 111:480-487.

Mukku, V.R., and Stancel, G.M. 1985a. Regulation of uterine

epidermal growth factor receptors by estrogen. J. Biol. Chem. 260:9820-9824.

Mukku, V.R., and Stancel, G.M. 1985b. Receptors for epidermal growth factor in the rat uterus. Endocrinology 117:149-154.

Murmatsu, I; Hollenberg, M.D.; Lederis, K. 1985. Vascular actions of epidermal growth factor-urogastrone: possible relationship to prostaglandin production. Can. J. Physiol. Pharmacol. 63:994-999.

Murphy, L.J., Murphy, L.C., and Friesen, H.G., 1987a. Estrogen induces insulin-like growth factor-1 expression in the rat uterus. Mol. Endocrinol. 1:445-450.

Murphy, L.J., Murphy, L.C., and Friesen, H.G. 1987b. Estrogen induction of N-myc and c-myc protooncogene expression in the rat uterus. Endocrinology 120:1882-1888.

Murphy, L.J., and Friesen, H.G. 1988. Differential effects of estrogen and growth hormone on uterine and hepatic insulin-like growth factor I gene expression in the ovariectomized hypophysectomized rat. Endocrinology 122:325-332.

Nishikura K, and Murray J.M. 1987. Antisense RNA of c-fos blocks renewed growth of quiesent 3T3 cells. Mol. Cell. Biol. 7:636-649.

Rauscher, F.J. III, Sambucetti, L.C., Curran, T., Distel, R.J., and Speigleman, B.M. 1988. A common DNA binding site for fos protein and transcription factor AP-1. Cell 52:471-480.

Ruther, U., Garber, C., Komitowski, D., Muller, R., and Wagner, E.F. 1987. Deregulated c-fos expression interferes with normal bone development in transgenic mice. Nature 325:412-416.

Stiles, C.D., Capone, G.T., Scher, C.D., Antoniades, H.N., Van Wyk, J.J., and Pledger, W.J. 1979. Dual control of cell growth by somatomedin and platelet-derived growth factor. Proc. Natl. Acad. Sci. 76:1279-1283.

Tomooka, Y., DiAugustine, R.P., McLachlan, J.A. 1986. Proliferation of mouse uterine epithelial cells in vitro. Endocrinology 118:1011-1018.

Travers, M.T., and Knowler, J.T. 1987. Oestrogen - induced expression of oncogenes in the immature rat uterus. FEBS LETT. 211:27-30.

Treisman, R. 1985. Transient accumulation of c-fos RNA following serum stimulation requires a conserved 5' element and c-fos 3' sequences. Cell 42:889-902.

Verma, I., and Sassone-Corsi, P. 1987. Protooncogene fos: complex but versatile regulation. Cell 51:513-514.

Weisz, A., and Bresciani, F. 1988. Estrogen induces expression of c-fos and c-myc protooncogenes in rat uterus. Mol. Endocrinol. 2:816-824.

Current <u>Addresses</u>: V.R.M. - Genetch, Inc., 460 Point San Bruno Blvd., South San Francisco, CA 94080; R.B.L. -Merck and Co. 80Y-310, Rahway, NJ 07065; R.M.G. - Dept of Biology, Villanova University, Villanova, PA 19085.

DISCUSSION OF THE PAPER PRESENTED BY G. STANCEL

KORACH: Have you looked at stimulation of c-fos with 16 alpha estradiol?

STANCEL: We have not yet completed these experiments. I agree, however, that it will be very interesting to determine how short acting estrogens like the 16 alpha compound affect fos induction and other parameters. On the other hand, we have studied EGF receptor induction by 16 alpha estradiol. A single injection does not induce EGF receptor synthesis, but multiple injections do. In other words, the pattern of EGF receptor synthesis in response to 16 alpha estradiol is the same as the pattern of DNA synthesis.

TATA: Does EGF-R gene have an ERE?

STANCEL: The sequence data available does not reveal an ERE in the upstream region of the EGF receptor gene. However, we have no way of knowing how far upstream (or downstream) such an element might be, i.e., it is possible that an ERE will turn up as more upstream regions of the EGF receptor gene are sequenced.

TATA: I would like to make the comment that in estrogen targets in which the hormone does not stimulate DNA synthesis, as in induction of vitellogenin mRNA in Xenopus hepatocytes, it reduced the amount of EGF-R. This is compatible with an earlier finding that inhibition of DNA synthesis by hydroxyurea produced a potentiation of the hormonal induction of vitellogenin synthesis.

THOMPSON: Does c-fos product negatively regulate its own gene?

STANCEL: That's possible, but I don't think the question of whether fos protein directly regulates its gene in a negative manner is established either way. There is some evidence that AP-1, or maybe a complex of AP-1 and fos, negatively controls fos expressions. There is also evidence that fos is regulated by repression in resting cells, and this would presumably involve other regulatory molecules.

LIAO: In the "double hit" experiment, when $E_z$ effect was seen after/during the rate of increase in DNA synthesis was seen, what proportion of cells was actually dividing? In many organs increase in the rate of DNA synthesis in response to steroid hormones are due to small number of total cells. If so, one cannot conclude that second $E_z$ hit require cell division.

STANCEL: We did not measure the proportion of cells dividing in the same animals in which we did the double-hit studies. However, in similar studies with this model system (i.e., the 20 day old animal) we have measured that by continuous labelling with tritiated thymidine. The results of these studies indicate conclusively that the majority of cells, at least epithelial cells, do enter S phage and undergo DNA replication.

ROY: In your "early" and "late" effect model I wondered why you did not try to count these with tyrosine kinase? Also have you examined the effect of tamoxifen on the induction of fos/myc induction?

STANCEL: The EGF receptor contains a tyrosine kinase activity which is stimulated by growth factor binding. This would clearly be one point in our model where tyrosine phosphorylation would come into play. Unfortunately, it has been very difficult to identify tyrosine kinase substrates which have an established role in cell growth. It is also possible that other receptors may be involved in tyrosine phos-phorylation. For example, Murphy and Friesen and their colleagues have evidence that IGF-1 might be involved in estrogen action, and that cascade might involve activation of a tyrosine kinase. In response to your second question, we are in the process of examining effects of antiestrogens on fos and myc expression, but these studies are not yet completed.

HARDIN: Have you looked at fos protein levels following $E_2$ injection?

STANCEL: The results with fos expression are very recent, and we simply have not had the time to study fos protein levels yet. These will be very important studies, however, because the fos system is probably regulated at every conceivable level, and will likely be regulated by both positive and negative control mechanisms.

HARDIN: What is the time course of fos mRNA levels following puromycin treatment?

STANCEL: The time courses of fos mRNA induction are generally similar following estrogen alone or estrogen plus puromycin, although we don't have quite as many time points for the samples with the puromycin. We really have not studied the time course of puromycin effects alone.

DISCUSSANTS: K. KORACH, G. STANCEL, J.R. TATA, E.B. THOMPSON, S. LIAO, A.K. ROY AND J. HARDIN.

# Regulation of the Rat Insulin II Gene: *cis-* and *trans-acting* Factors

Young-Ping Hwung, David T. Crowe, Lee-Ho Wang, Sophia Y. Tsai
and Ming-Jer Tsai

Department of Cell Biology
Baylor College of Medicine
One Baylor Plaza
Houston, TX 77030

## Introduction

Insulin is one of the key peptide hormones in the control of growth
and metabolism.  The imbalance of insulin action results in diabetes
mellitus which has broad impact on the society.  Aside from its
medical importance, the insulin gene is interesting because it is
specifically expressed in the β-cells of the islets of Langerhans in
the pancreas, and this tissue-specificity is at the transcriptional
level.  Therefore, the insulin gene serves as a good model system for
studies of gene expression.

The molecular basis of tissue-specific expression of the insulin gene
has been partially defined.  Hybrid genes containing varying lengths
of insulin 5'-flanking sequences joined to the coding region of the
bacterial chloramphenicol acetyl transferase (CAT) gene were
transfected into different cell lines.  The results demonstrated that
302 base pairs of 5'-flanking sequence from the rat insulin I gene
(rInsI) are sufficient to direct CAT expression in insulin-producing
cells, but not in non-insulin producing cells (Walker *et al.*).

Subsequently, it was found that two distinct sequence elements, the enhancer and promoter, reside in this region, and each of them can confer tissue-specificity (Edlund *et al.*). Similar results were obtained with the rat insulin II gene (rInsII), which is highly homologous to and equally expressed as rInsI. rInsII sequences were fused to the selectable marker xanthine/guanine phosphoribosyl-transferase gene (gpt) and the 5'-flanking sequences led to a relatively high frequency of resistant cells only in insulin-producing cells (Episkopou *et al.*). In addition, introduction of rInsII-SV40 T antigen fusion genes into transgenic mice resulted exclusively in pancreatic β-cell tumors (Hanahan).

Negative sequence elements may play a role to inhibit insulin gene expression in inappropriate cell types. Sequences 2 to 4 kilobases upstream from the rInsI transcription start site were shown to have silencer activity (Laimins *et al.*). The addition of large excess of sequences within the rInsI enhancer ( -249 to +50) were also found to mediate de-repression of a rInsI-enhancer driven gene activity in cells that do not produce insulin (Nir *et al.*).

All the observations described above suggest that tissue-specificity of the insulin gene can be achieved at the level of transcriptional initiation, and the interaction between *trans-acting* factors and the 5'-flanking region of the insulin gene may be important. To further understand the molecular mechanism, we determined the sequence elements in greater detail, and characterized some *trans-acting* factors binding to the important sequence elements.


**Results and Discussion**


*Cis-acting* elements


The 5'-flanking sequences of rInsII were deleted to different end-points to determine the minimal sequence requirement for efficient expression of rInsII in a hamster insulinoma cell line, HIT cells. rInsII sequences were fused to the CAT gene coding region, transiently transfected into HIT cells and CAT activities were determined. Deletion of 5'-flanking sequence to -218 (relative to the cap site) does not change the level of expression when compared to the wild-type control plasmid which contains the full length enhancer and promoter (up to -448). However, further deletions from -218 to -179 and -179 to -162 each reduced the activity by at least 3 fold.

A series of linker scanning (LS) mutants were constructed to better define the important sequences without changing the spacing and organization of individual sequence motifs. Five to twenty base pairs of wild-type sequences in the 5'-flanking region of rInsII were replaced by a SacI linker in each LS mutant, and the coding region of the CAT gene is linked as a reporter. Figure 1 indicates that some LS mutants displayed similar or even higher activity compared to the wild-type plasmid, therefore the Sac I linker is not detrimental to the expression of rInsII. Disruption of the sequences from -128 to -111 or -101 to -93 decreased the activity to 4% of that of the wild-type plasmid, suggesting that these two regions are indispensable for the expression of rInsII. Sequences from −53 to −46 and −172 to −163 are also important, since mutation of either region reduced expression more than 80%. Several other mutations (−298 to −289, −227 to −219, −189 to −180, and −145 to −137) reduced activity to 30−50%, so they represent moderately important sequences in the 5'-flanking region.The important sequence elements in the 5'-flanking region of rInsI, the homologous gene of rInsII, has been determined with a similar approach (Karlsson et al.). rInsI and rInsII have a sequence similarity of 85% in the 500 base pair sequences immediately upstream from the transcription start site, and they are similarly regulated under normal physiological conditions (Clark and Steiner, Giddings and Carnaghi). Thus the cis-acting elements in these two genes were expected to be very similar. However, efficient expression of rInsII needs only 218 base pairs of 5'-flanking sequence, whereas the 5'-border of the rInsI enhancer was mapped to −302, and the deletion from −302 to −219 resulted in a 100-fold decrease in activity. In addition, there are two homologous sequence elements in rInsI, −241 to −233 and −112 to −104, which are essential

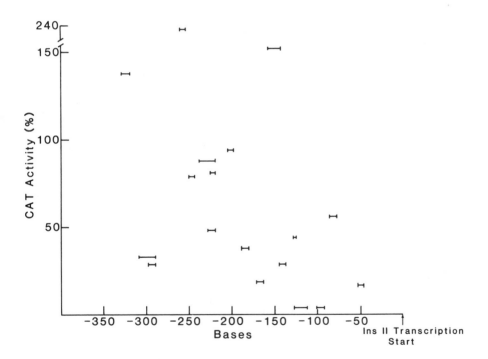

Figure 1. Detection of important rInsII sequence elements with linker-scanning mutants. The activity of each linker-scanning mutant is indicated as percentage of the activity of the wild-type plasmid which contains the full length rInsII enhancer and promoter. The X-axis shows the mutated regions relative to the transcription start site.

for the expression. These two elements are similar to the immunoglobulin heavy chain μ concensus sequence, and they bind specific *trans-acting* factors (Moss *et al.*). Surprisingly, the corresponding upstream region in rInsII can be mutated without any change in activity. Finally, several elements essential for rInsII are not found to be important for rInsI.

rInsI is hypothesized to be originated from a retroposome-mediated gene duplication of rInsII (Soares *et al.*), while most other mammals have only one insulin gene which has the same gene organization as rInsII (Steiner *et al.*). This is consistent with the fact that the results of deletion analysis of the human insulin gene are more similar to that of rInsII, rather than rInsI. Therefore, rInsI may have acquired new sequence elements during evolution to compensate for the lost sequence elements. Another possibility is that different sequence elements of rInsI and rInsII may enable them to respond to some undetermined regulators in different manners.

*Trans-acting* factors

The presence of multiple sequence elements in the 5'-flanking region of the rat insulin I gene is not surprising. More and more enhancers and promoters have been shown to possess multiple sequence motifs, which may bind DNA-binding proteins and communicate with each other through the interaction between their cognate binding proteins. It will be essential to isolate and characterize the *trans-acting* factors before the molecular basis of enhancer and promoter functions can be fully understood.

Binding factors to some upstream sequence elements in the 5'-flanking region were detected with the band-shifting assay. Oligonucleotides containing the important sequences as defined by the LS mutants were synthesized and used as probes to assay for DNA-binding proteins in the nuclear extract from HIT cells. Each of four probes (-305 to -281, -193 to -159, -153 to -129 and -125 to -86) showed specific binding complexes which were not competed by a large excess of the non-specific competitor, poly d(I-C), or the other three oligonucleotides. Nevertheless, they could be competed by the homologous oligonucleotide. Therefore there are sequence-specific binding factors in HIT cells for these important elements. The tissue distribution and the characteristics of these factors are being investigated.

An interesting binding activity involves in LS-87/76 in which sequences from -86 to -77 has been replaced by a Sac I linker. When the rInsII promoter was used in band-shifting assays, a protein-DNA complex appeared only when HIT cell nuclear extracts were used, but not in HeLa, B lymphocytes, liver and several other cell types. When the probe contained the LS-87/76 mutation, the binding complex disappeared, indicating that natural sequences from -86 to -77 are required for the binding of this HIT cell-specific factor. Since the insulin promoter is capable of directing tissue-specific expression, this HIT cell-specific promoter-binding factor is a potential candidate to a tissue-specific transcription factor. However, LS-87/76 displayed a 56% activity compared to the wild-type plasmid, suggesting that this region is not absolutely required for the expression of rInsII. One explanation to these seemingly contradicting results is that multiple redundant tissue-specific elements exist within the enhancer and promoter regions, and deleting one of these elements may not exert a major effect on the tissue-specificity.

The other promoter mutant LS-54/45 has only 15% of the wild-type activity in transfection experiments, therefore the region being mutated (-53 to -46) defines an important promoter element, called RIPE (rat insulin promoter element). The binding protein to RIPE has been well characterized and turned out to be very interesting. The binding factor could be detected in both HIT cells and non-insulin producing cells (e.g. HeLa cells) by the band-shifting assay. The binding region spans from -60 to -40, as determined by DNase I footprinting analysis. Mutation of this region destroyed both the *in vivo* gene activity and the *in vitro* binding activity, indicating that the RIPE-binding protein may play a functional role in the expression of rInsII.

The binding protein was purified from HeLa cells. Surprisingly, it is the same protein as the COUP (chicken ovalbumin upstream promoter) transcription factor which binds to the upstream promoter of the chicken ovalbumin gene and is required for its efficient transcription (Pastorcic et al, Sagami *et al.*, Wang *et al.*). There are several lines of evidence showing that the COUP transcription factor binds to RIPE (Hwung *et al.*, 1988a,b). First, the two binding activities copurified through several different chromatographic columns, including the final sequence-specific affinity column containing the COUP oligonucleotide. Second, the insulin promoter can compete with the ovalbumin promoter for the COUP transcription factor, and similarly, the ovalbumin promoter can compete with the insulin promoter for the RIPE binding factor. Third, the affinity-column purified COUP transcription factor was resolved in a SDS-polyacrylamide gel and the polypeptides were eluted and renatured. Each polypeptide which binds to the ovalbumin promoter also binds to the insulin promoter. Finally, a series of oligonucleotides harboring single or double point mutations in the COUP sequences were used to compete with either promoter for the corresponding binding factor. The mutants which were unable to bind to the COUP transcription factor also failed to bind to the RIPE binding factor, and vice versa.

The binding of the COUP transcription factor to these two promoters was unexpected because the two binding sequences do not have obvious similarity. Moreover, the GC contents of the COUP and the RIPE sequences are 36% and 75%, respectively, suggesting that the rigidity of these two regions may also be very different. What are the important sequences for the recognition and binding of the COUP transcription factor? How does this factor bind to the two different sequences? To determine the sequences required for the binding of the transcription factor, we synthesized a series of oligonucleotides which bear single or double point mutations in the COUP sequence. These oligonucleotides were used to compete for the binding of the COUP transcription factor with either the ovalbumin or the insulin promoter. Figure 2 shows that mutation at G-77, G-76, or the simultaneous mutation of C-81 and C-74 abolished the binding activity. Mutation of C-74, A-78, A-79, A-80, C-81 or G-83 reduced binding activity, while G-85 can be mutated without any effect. The same results were obtained with both promoters.

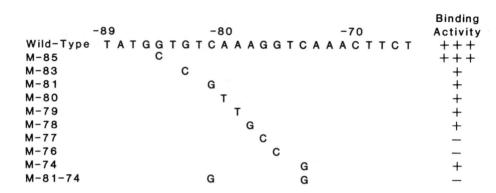

Figure 2. Binding affinities of mutated COUP oligonucleotides. Sequence of the wild-type COUP oligonucleotide is shown on top. Each mutated oligonucleotide has a single or double point mutation as indicated.

The purine and phosphate groups which contact with the COUP transcription factor were determined by methylation and ethylation interference assays, respectively, for both promoters. The results were taken together with the mutation data described above to deduce the best alignment of the two binding sequences (Figure 3). This alignment maximizes the similarity in the distribution of purine and phosphate contact points, while the important nucleotides are aligned

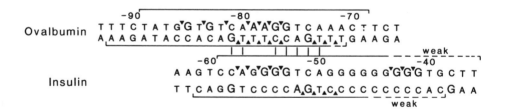

Figure 3. Sequence comparison of the COUP transcription factor binding sites in the ovalbumin and insulin genes. The two sequences are aligned to maximize the similarity. The nucleotides shown in big letters are the purine contact points. Triangles are the phosphate contacts points. DNase I footprinted regions are indicated by the brackets. Vertical bars between the two sequences mark the common nucleotides.

as much as possible. While this alignment reveals a certain degree of similarity between the two binding sites, there is still substantial difference. In addition, the DNase I protected regions can not be aligned, which is unusual when the same binding factor is involved.

The purine and phosphate contacts data were analyzed with computer graphics. The computer-generated images indicate that the purine contacts are located on one side of the DNA helices in both sequences. However, the distribution of the phosphate contacts suggests that while the COUP transcription factor wraps around the ovalbumin promoter, it binds to only one face of the insulin promoter (Hwung et al., 1988b). Furthermore, the contact with the insulin promoter is more extended along the DNA helix than that with the ovalbumin promoter. Again these results indicate that the COUP transcription factor binds to the two promoters in different ways, although some similarities exist.

Several possibilities were addressed to explain how a specific DNA-binding protein binds to two different sequences in different ways. The protein may consist of two binding domains which have different binding specificities. These two domains are not able to bind DNA simultaneously, either due to steric hindrance or allosteric changes of one domain induced by the occupation of the other. If this is true, then the two binding sequences do not have to be similar. Alternatively, only one binding domain may be involved if it recognizes a sequence or structure with some flexibility. This possibility predicts that the binding sequences are similar to some extent. We proposed a working model which is a combination of the above two possibilities (Figure 4). The protein may have two overlapping DNA-binding domains. The overlapping region recognizes the common nucleotides in the two sequences, while other unique regions of the protein recognize the other nucleotides. This model is consistent with the similarities and the differences of the two binding sites, as well as non-alignment of the footprinted regions. We are currently in the process of cloning the COUP transcription factor, and the model will be tested with the cloned gene in the future.

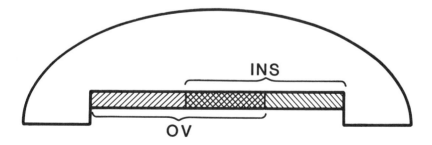

Figure 4. A working model for the DNA binding domains of the COUP transcription factor.OV and INS: the COUP transcription factor binding domains for the ovalbumin and insulin genes, respectively. The significance of the differential binding of the COUP

transcription factor to different genes is unknown. It may be another example of a factor binding to a variety of sequences with different affinities. The CCAAT-box binding factors are a group of proteins, each of which binds to a battery of related sequences with different affinities (Dorn et al., Chodosh et al.). Furthermore, the differential binding may facilitate distinct regulatory mechanisms. For example, the COUP transcription factor may assume different conformations when it binds to different genes, and thereby it may interact with different regulatory factors. It was shown that the COUP transcription factor interacts with a non-DNA binding protein, S300-II, to efficiently transcribe the chicken ovalbumin gene (Tsai et al.). Although the COUP transcription factor alone is unlikely to be responsible for tissue-specificity of the insulin promoter because it is not specific for insulin-producing cells, we do not exclude the possibility that it can interact with a pancreatic β-cell specific factor which is required for the tissue-specific insulin promoter function.

**Acknowledgement**

We thank T. Klein and R. Langridge of the University of California, San Franscisco, for computer graphics, J. Codina-Salada for oligonucleotide synthesis, and the members of the Tsai laboratory for helpful discussions. This work was supported by NIH grant HD-17379.

**Abbreviations**

rInsI: rat insulin I gene.
rInsII: rat insulin II gene.
COUP: chicken ovalbumin upstream promoter.
RIPE: rat insulin promoter element.
CAT: chloramphenicol acetyl transferase.
OV: ovalbumin gene.
INS: Insulin gene.
bp: base pair(s)

**References**

Chodosh, L.A., A.S. Baldwin, R.W. Carthew and P.A. Sharp (1988) Human CCAAT-binding proteins have heterologous subunits. Cell 53:11-24.

Clark, J.L. and D.F. Steiner (1969) Insulin biosynthesis in the rat: demonstration of two proinsulins. Proc. Natl. Acad. Sci. USA 62: 278-285.

Crowe, D.T. and M.-J. Tsai (1988) Mutagenesis of the rat insulin II 5'-flanking region defines sequences important for expression in HIT cells. Mol. Cell. Biol. Submitted.

Dorn, A., J. Bollekens, A. Staub, C. Benoist and D. Mathis (1987)   A multiplicity of CCAAT box-binding proteins. Cell 50: 863-872.

Edlund, T., M.D. Walker, P.J. Barr and W.J. Rutter (1985)   Cell-specific expression of the rat insulin gene:   Evidence for role of two distinct 5'-flanking elements. Science 230: 912-916.

Episkopou, V., A.J.M. Murphy and A. Efstratiadis (1984)   Cell-specified expression of a selectable hybrid gene. Proc. Natl. Acd. Sci. USA 81: 4657-4661.

Giddings, S.J. and L.R. Carnagni (1988)   The two nonallelic rat insulin mRNAs and pre-mRNAs are regulated coordinately *in vivo*. J. Biol. Chem. 263: 3845-3849.

Hanahan, D. (1985)   Heritable formation of pancreatic β-cell tumors in transgenic mice expressing recombinant insulin-simian virus 40 oncogenes. Nature 315: 115-122.

Hwung, Y.-P., D.T. Crowe, L.-H. Wang, S.Y. Tsai and M.-J. Tsai (1988a)   The COUP transcription factor binds to an upstream promoter element of the rat insulin II gene. Mol. Cell. Biol. 8: 2070-2077.

Hwung, Y.-P., L.-H. Wang, S.Y. Tsai and M.-J. Tsai (1988b) Differential binding of the COUP transcription factor to two different promoters. J. Biol. Chem. 263: 13470-13474.

Karlsson, O., T. Edlund, J.B. Moss, W.J. Rutter and M.D. Walker (1987)   A mutational analysis of the insulin gene transcription control region: Expression in beta cells is dependent on two related sequences within the enhancer. Proc. Natl. Acad. Sci. USA 84: 8819-8823.

Laimins, L., M. Holmgren-König and G. Knoury (1986) Transcriptional "silencer" element in rat repetitive sequences associated with the rat insulin I gene locus. Proc. Natl. Aca. Sci. USA 83: 3151-3155.

Moss, L.G., J.B. Moss and W.J. Rutter (1988)   Systematic binding analysis of the insulin gene transcription control region:   Insulin and immunoglobumin enhancers utilize similar transactivators.   Mol. Cell. Biol. 8: 2620-2627.

Nir, U., M.D. Walker and W.J. Rutter (1986)   Regulation of rat insulin I gene expression:   Evidence for negative regulation in nonpancreatic cells.  Proc. Natl. Acad. Sci. USA 83: 3180-3184.

Pastorcic, M., H. Wang, A. Elbrecht, S.Y. Tsai, M.-J. Tsai and B.W. O'Malley (1986)   Control of transcription initiation *in vitro* reguires binding of a transcription factor to the distal promoter of the ovalbumin gene. Mol. Cell. Biol. 6: 2784-2791.

Sagami, I., S.Y. Tsai, H. Wang, M.-J. Tsai and B.W. O'Malley (1986) Identification of two factors required for transcription of the ovalbumin gene. Mol. Cell Biol. 6: 4259-4267.

Soares, M.B., E. Schon, A. Henderson, S.K. Karathanasis, R. Cate, S. Zeitlin, J. Chirgwin and A. Efstratiadis (1985)   RNA-mediated gene duplication:   The rat preproinsulin I gene is a functional retroposon. Mol. Cell. Biol. 5: 2090-2103.

Steiner, D.F., S.J. Chan, J.M. Welsh and S.C.M. Kwok (1985) Structure and evolution of the insulin gene. Ann. Rev. Genet. 9: 463-84.

Tsai, S.Y., I. Sagami, H. Wang, M.-J. Tsai and B.W. O'Malley (1987) Interactions between a DNA-binding transcription factor (COUP) and a non-DNA binding factor (S300-II). Cell 50: 701-709.

Walker, M.D., T. Edlund, A.M. Boulet and W.J. Rutter (1983) Cell-specific expression controlled by the 5'-flanking region of insulin and chymotrypsin genes. Nature 306: 557-561.

Wang, L.-H., S.Y. Tsai, I. Sagami, M.-J. Tsai and B.W. O'Malley (1987) Purification and characterization of chicken ovalbumin upstream promoter transcription factor from HeLa cells. J. Biol. Chem. 262: 16080-16086.

DISCUSSION OF THE PAPER PRESENTED BY M. TSAI

ROY: Ming - when you discovered that COUP binds to Insulin promoter you must have examined many other promoters. If you have found any other binding what are the significance with respect to common regulatory process in these COUP binding genes?

TSAI: Yes, we have examined a few gene promoters. The adenosine deaminase, SV40 early, MMTV, lysozyme, $\beta$-globin and Herpes Simplex virus TK gene promoters do not have detectable binding site. While the ovalbumin, Y (ovalbumin related), ovomucoid, insulin, and maybe VLDL genes contain a binding site for the COUP transcription factor. Since the binding sites exist in such diverse gene promoters, it is not likely that these genes share a common regulatory process.

SIMONS: The receptor fragment that you have used (AA 394-500) is larger than the minimal DNA binding domain. Have you looked to see if you can remove some of the "excess" amino acids and separate the ability to bind to DNA from the ability to exhibit cooperativity in the DNA binding:

TSAI: This is a very good question. Currently, we are mutating the receptor gene in the hope to identify the sequence important for the cooperativity.

TCHEN: With reference to cooperativity - is there indication of leucine zippers?

TSAI: Yes, there is a potential leucine zipper in the hormone binding domain of human and chicken progesterone, human estrogen, human glucocorticoid Vitamin D and thyroxine receptors. In addition, we find another potential leucine zipper in the $T_1$ region of the chick progesterone receptor. The potential role of the leucine zipper in dimerization of hormone receptor is not known at present.

DISCUSSANTS: A.K. ROY, M. TSAI, S.S. SIMONS, AND T. TCHEN

# Thyroid Hormone Regulation of Rat Liver S14 Gene Expression.

Donald B. Jump

## Introduction

Thyroid hormones have diverse effects on a wide range of physiological and biochemical processes such as endocrine function, metabolism, and growth and development (Wolff and Wolff, 1969). The active form of the hormone, triiodothyronine (T$_3$), mediates changes in cellular function by binding to limited capacity high affinity receptors located in nuclei of target cells (for review see Oppenheimer et al., 1987). In contrast to steroid hormone action, the association of T$_3$ receptors with chromatin is not dependent on the presence of the hormone (Dillman et al., 1974; Surks et al., 1975). T$_3$ receptors function to regulate the transcriptional activity of a limited number of genes leading to specific changes in phenotypic expression. For example, T$_3$ stimulates the transcription of the rat pituitary growth hormone gene (Yaffe and Samuels, 1984), hepatic malic enzyme (Dozin et al., 1986) and HMG-CoA-reductase genes (Simonet and Ness, 1988), while inhibiting the transcription of thyroid-stimulating hormone gene (Shupnik et al., 1985) and the expression of several members of the myosin heavy chain gene family (Izumo et al., 1986).

Recent studies show the cellular erbA proto-oncogene product is related to the thyroid hormone receptor (Sap et al., 1986; Weinberger et al., 1986). This is based on the finding that c-erbA polypeptides expressed by in vitro transcription-translation from cloned cDNA's bind thyroid hormones with the affinity and analog specificity as found for the native T$_3$ receptor. The molecular weight of the in vitro synthesized proteins, designated c-erb-$\alpha$ (46 kd) and c-erb -$\beta$ (52 kd) is similar to the T$_3$ native receptor. Sequence analysis of the c-erbA genes show these proteins are related to steroid receptors (Weinberger et al., 1986; Sap et al., 1986). Since both receptor types function as regulators of gene expression, some investigators have suggested steroid and thyroid hormone receptors are members of a family of nuclear proteins functioning as "ligand-inducible transcription factors" (Evans, 1988). A considerable body of evidence is available to support such a role for steroid receptors (Yamamoto, 1985; Butti and Kuhnel, 1986; Godowski et al., 1988) and recent studies on the T$_3$ regulation of rat pituitary growth hormone synthesis also support this notion.

Induction of growth hormone gene transcription by T$_3$ is directly related to receptor occupancy (Yaffe and Samuels, 1984). Both a T$_3$ receptor binding site (Glass et al., 1987; Evans, 1988) and sequences essential to $T_3$ regulation of growth hormone transcription are located proximal to the growth hormone gene promoter (-158 to -194 bp) (Flug et al., 1987: Glass et al., 1987; Koenig et al., 1987; Wight et al., 1987, 1988; Ye et al., 1988). Thus, a thyroid hormone-responsive element (TRE) associated with the growth hormone gene appears to function in a fashion similar to the well characterized glucocorticoid-responsive elements (GRE's) associated with the long-terminal repeat of murine mammary tumor virus (Yamamoto, 1985; Butti and Kuhnel, 1986) and the hepatic tyrosine aminotransferase gene (Jantzen et al., 1987).

Despite our current understanding of the structure of thyroid hormone receptors, surprisingly little information is known about how $T_3$ receptors function at the chromatin level to induce changes in gene transcription. Our approach to this problem has focused on the $T_3$-mediated regulation of the transcriptional activity and chromatin structure of the rat liver S14 gene. I will discuss the complex regulation of S14 gene expression by tissue-specific, developmental, hormonal and nutritional factors and provide evidence to suggest $T_3$ initiates changes in S14 gene transcription by inducing a site specific modification of chromatin structure upstream from the S14 gene.

## The S14 Model

The S14 protein was first described in studies designed to examine the diversity of $T_3$ effects on hepatic gene expression (Seelig et al., 1981). Hepatic mRNA isolated from hypothyroid, euthyroid, and hyperthyroid animals programmed the synthesis of [$^{35}$S]methionine-labeled proteins in an in vitro translation system. The labeled proteins were separated by 2-dimensional gel electrophoresis and their distribution detected by autoradiography. The technique provided an indirect measure of the effects of $T_3$ on hepatic gene expression at the pre-translational level. Of the nearly 250 translated products resolved by the two-dimensional gel analysis, 8% were affected by thyroidal status. While several mRNA's were induced in the transition from hypothyroidism to hyperthyroidism, others were repressed, indicating that $T_3$ had both positive and negative effects on hepatic gene expression. One of the translated products induced significantly by increasing plasma $T_3$ levels was "Spot #14" or S14. The $mRNA_{S14}$ coded for a protein of 17,000 $M_r$ and 4.9 pI. Subsequent studies showed that the $T_3$-mediated induction of $mRNA_{S14}$ represented one of the earliest responses of the liver to thyroid hormone (Seelig et al., 1982).

These studies suggested that the regulation of $mRNA_{S14}$ expression may be a good model to examine the rapid effects of $T_3$ on hepatic gene expression. In subsequent studies, Towle and colleagues cloned the S14 cDNA's and gene coding for the S14 protein and found the S14 protein was coded from a single copy gene of 4.4 kb containing two exons (Liaw and Towle, 1984). Two polyadenylation signals located in the 3'exon explain the presence of two $mRNA_{S14}$ of 1.3 and 1.47 kb detected on northern analysis (Liaw and Towle, 1984; Narayan et al., 1984; Jump et al., 1984). DNA sequence analysis of the S14 cDNA's showed a single open reading frame in the 5'exon coding for a protein of 17,010 $M_r$. This value agreed favorably with the size of the in vitro translated product and a $T_3$-responsive protein identified in hepatic cytosol (Jump et al., 1984). Comparison of the S14 DNA and protein sequence with sequences in the national databases, however, failed to reveal significant homologies with known proteins. Although the precise biochemical function the S14 protein is unknown, studies on the tissue distribution of the protein and the regulation of expression of mRNAs$_{14}$ by specific physiological stimuli suggests the S14 protein functions in some aspect of lipid metabolism (Jump et al., 1984; Jump and Oppenheimer, 1985; Jump et al., 1986; Freake and Oppenheimer, 1987).

## Tissue-Specific Regulation of S14 Gene Expression

The tissue distribution of S14 expression was examined to gain additional insight into the function of the S14 protein and to determine whether S14 gene expression was under tissue-specific control. The relative level of mRNAs$_{14}$ expression in various rat tissues was measured by dot blot and northern analysis using cDNAs$_{14}$ probes (Table I). The results were normalized to levels of mRNAs$_{14}$ expression in a rat liver standard to aid in the comparison of the relative abundance of mRNAs$_{14}$ in various rat tissues. The relative level of mRNAs$_{14}$ expression in lactating mammary gland and epididymal fat was 2-fold and 6-fold higher, respectively, than in rat liver. Whereas the relative level of mRNAs$_{14}$ expression in lung and brain was $\leq$ 6% of the liver value, the level in spleen, kidney, heart and skeletal muscle was $\leq$ 1.0 % of the level in rat liver. Northern analysis showed that [$^{32}$P]cDNAs$_{14}$ hybridized to two mRNA species with identical electrophoretic characteristics in each tissue, i.e., 1.3 and 1.47 kb (Jump et al., 1984; Jump and Oppenheimer, 1985).

The high level of mRNAs$_{14}$ expression in white adipose tissue suggests the S14 protein may participate in lipid metabolism. However, the fact that liver, lactating mammary gland and white adipose tissue all function

Table 1: Tissue-Specific Regulation of S14 Gene Expression.

| Tissue | Relative level of mRNAs$_{14}$ [a] | In Vitro Transcriptional S14 Run-On Activity (ppm) [b] |
| --- | --- | --- |
| Liver | 1.0 + 0.3 | 36.2 + 2.0 |
| Mammary Gland | 2.1 + 0.6 | – |
| Epididymal Fat | 6.0 + 1.0 | – |
| Heart | 0.005 + 0.002 | – |
| Skeletal Muscle | 0.002 + 0.001 | – |
| Lung | 0.020 + 0.005 | – |
| Brain | 0.060 + 0.010 | <0.1 |
| Kidney | 0.009 + 0.005 | <0.1 |
| Spleen | 0.003 + 0.001 | <0.1 |

[a]Relative level of mRNAs$_{14}$ was measured by dot blot analysis and the abundance of S14 expression compared to an internal RNA hybridization standard derived from the liver of a male rat (Jump and Oppenheimer, 1985). The results are reported as mean + S.D. with 4 samples in each group. The relative levels of mRNAs$_{14}$ expression in the various tissues was similar to previously reported values (Jump and Oppenheimer, 1985).
[b]The in vitro transcriptional run-on activity of the S14 gene has been described (Jump. et al., 1988). This assay measured S14 run-on activity in nuclei isolated from liver, brain, kidney and spleen of 250 g. male rats. The values are expressed as mean + S.D. for liver (N=20). The S14 transcriptional activity in brain, kidney and spleen (N=2 for each tissue) was below the level of detection. (–) Not measured.

in lipid synthesis, storage, and metabolism (Nicols and Locke, 1984; Gardener et al., 1983) makes it impossible to assign a precise function to the S14 protein. Moreover, examination of the enzymes involved in fat metabolism (Brindley, 1985; Cook, 1985; Goodridge, 1985) and intra-cellular fatty acid transport (Alpers et al., 1984) does not point to any likely candidate possessing the biochemical or physiological characteristics of the S14 protein. Thus, the S14 protein may be a novel protein involved in lipid metabolism (Jump and Oppenheimer, 1985).

**Tissue-Specific Regulation of S14 Gene Transcription.**

Our studies have focused on the transcriptional regulation of S14 gene expression. The transcriptional activity of the S14 gene was measured using an _in vitro_ nuclear transcriptional run-on assay which was modified from assays developed by Clayton and Darnell (1983) and Shupnik et al., (1985). Briefly, isolated nuclei are incubated with [$^{32}$P]-UTP to label transcripts that were initiated _in vivo_. The labeled RNA was isolated and hybridized to blots containing three plasmids homologous to distinct regions of the S14 gene. These plasmids contain sequences homologous to the 5'exon, a non-reiterated region of the intron and the 3'exon. The gene was considered transcriptionally active when [$^{32}$P]-RNA synthesized in isolated nuclei hybridized to each probe (Jump et al., 1988).

Whereas the S14 transcriptional activity in the euthyroid liver was $36.2 \pm 2.0$ ppm, the transcriptional activity of the S14 gene in nuclei from brain, kidney or spleen was below the level of detection. The major difference in mRNAs$_{14}$ abundance between liver, spleen, brain and kidney can therefore be attributed to the transcriptional activity of the S14 gene. Thus, the relative abundance of mRNAs$_{14}$ in the various rat tissues is controlled at the level of gene transcription.

**Chromatin Organization of the S14 gene**

In an effort to understand the molecular basis of the tissue-specific regulation of S14 gene expression, we examined the chromatin organization of the S14 gene. Several features distinguish transcriptionally active from inactive chromatin (for review see Weisbrod, 1985). In addition to the general preferential sensitivity of active genes to DNase I (Weintraub and Groudine, 1976), active chromatin is punctuated by local discontinuities in the nucleosomal array to form sites that are hypersensitive to DNase I (Elgin, 1981). DNase I hypersensitive sites are thought to arise from the interaction of DNA-binding proteins (trans-acting) with underlying DNA sequences (for review see Gross and Garrard, 1988). These modified chromatin structures are found frequently flanking the 5'end of active genes associated with promoter, enhancer and silencer elements.

Inducible DNase I hypersensitive sites flank the 5'end of genes regulated by glucocorticoids (Zaret and Yamamoto, 1984; Jantzen et al., 1987), progesterone (Hecht et al., 1988), thyroid hormone (Nyborg and Spindler, 1986; Usala et al., 1988) and lipopolysaccharide (Parslow and Granner, 1982,1983). The lipopolysaccharide-inducible hypersensitive site was found within the immunoglobulin gene enhancer of mouse pre-B-cells (Parslow and Granner, 1982,1983). The glucocorticoid-inducible sites within the long terminal repeat of murine mammary tumor virus (Zaret and Yamamoto, 1984) or upstream from the tyrosine aminotransferase gene (Jantzen et al., 1987; Schutz et al., 1987) and the progesterone-inducible sites upstream from the chicken oviduct lysozyme gene (Hecht et al., 1988) harbor hormone-responsive elements (HRE). The HRE's are inducible enhancers which serve as targets for binding ligand-activated steroid receptors to chromatin. Receptor binding induces changes in the transcription rate of cis-linked genes. These studies suggest that inducible DNase I hypersensitive sites play a key role in the hormonal

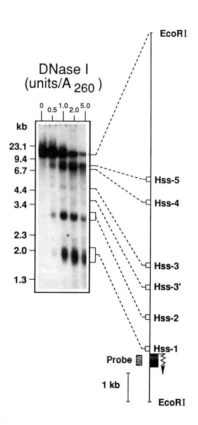

**Figure 1: DNase I Hypersensitive Sites Upstream from the Rat Liver S14 Gene.** Hepatic nuclei isolated from an adult male euthyroid rat were digested with varying concentrations of DNase I. The DNA was isolated and digested with EcoRI and electrophoretically separated in 1% agarose gels. The DNA was depurinated, transferred to Zetabind and hybridized with a 5'exon probe. DNA size markers were lambda DNA digested with HindIII and ϕX-174 DNA digested with HaeIII. The thin vertical line represents the 13 kb EcoRI fragment containing the 5'end of the S14 gene and contiguous flanking DNA. The 5'exon is represented by a black box with an arrow designating the location of the cap site and direction of transcription. The locations of the 6 DNase I hypersensitive sites are labeled and discussed in the text.

regulation of gene expression. Our goal has been to characterize the regulation of S14 DNase I hypersensitive sites by tissue-specific, developmental and hormonal factors.

In order to determine whether DNase I hypersensitive sites flank the S14 gene, the chromatin structure of the S14 gene was examined using the indirect-end labeling technique originally described by Wu (1980). The application of this technique to the S14 gene has been described (Jump et al., 1987; Jump et al., 1988). The gene map in Figure 1 represents only the 5'end of the S14 gene. The EcoRI site in the intron is 1.7 kb from the 5'end of the S14 gene while the distal EcoRI site is 11.3 kb upstream from the S14 cap site. Six major DNase I hypersensitive sites flank the 5'end of the S14 gene in the adult rat liver. No DNase I hypersensitive sites were found either within or flanking the 3'end of the S14 gene. No DNase I hypersensitive sites were detected when isolated liver DNA was digested with DNase I and subsequently digested with EcoRI indicating that the organization of the DNA in chromatin rather than the DNA sequences alone confer preferential DNase I sensitivity.

Figure 1 illustrates the pattern of DNase I hypersensitive sites located upstream from the 5'end of the S14 gene in liver nuclei of the euthyroid rat. The location of the 4 major and 2 minor DNase I hypersensitive sites relative to the 5'end of the S14 gene is: Hss-1 at -65 to -265 bp; Hss-2 at -1.3 kb; Hss-3' at -2.1 kb; Hss-3 from -2.8 to -3.3 kb; Hss-4 at -5.3 kb and Hss-5 at -6.5 kb. Whereas six DNase I hypersensitive sites flank the S14 gene in rat liver and lactating mammary gland, the Hss-1, 2, 3', and 3 hypersensitive sites were absent in brain, kidney and spleen. While Hss-4 and Hss-5 are constitutive DNase I hypersensitive sites, the presence of the tissue-specific DNase I hypersensitive sites Hss-1, 2, 3' and 3 correlates with the tissue-specific transcription of the S14 gene (Jump et al., 1987).

## Tissue-Specific Nuclear Proteins Interact with DNA Sequences Flanking the 5' End of the S14 Gene

Studies on the tissue-specific expression of the S14 gene were extended to the nuclear protein level to determine whether the formation of the Hss-1 site was due to the presence of proteins subject to tissue-specific expression. The Hss-1 site extends from -65 to -265 bp upstream from the S14 cap site (Jump et al., 1988). This structure is composed of two DNase I hypersensitive sites. A major site (Hss-1A) spans sequences from -65 to -180 bp, while a minor site (Hss-1B) spans sequences from -215 to -265 bp.

The presence of two adjacent sites separate by a region of relative DNase I resistance suggest DNA binding proteins interact with this region. To address this question, a gel shift analysis was used to detect the presence of sequence specific DNA binding proteins in rat liver nuclear extracts. The technique was based on the observation that DNA-protein complexes display decreased electrophoretic mobility compared to the unbound DNA fragments resulting from a change in size and/or charge. This approach was originally developed to examine the interaction of purified procaryotic regulatory proteins with specific DNA sequences (Garner and Revzin, 1981; Fried and Crothers, 1981).

In the present study, a 487 bp fragment extending from -462 to +25 bp relative to the S14 cap site was isolated from an S14 genomic clone (pS14EXOT1-8cc), digested with PstI (cuts at -290 bp) and AhaII (cuts at

-61 bp) to give three restriction fragments: "A" (229 bp), "B" (171 bp) and fragment "C" (87 bp). The fragments were end labeled with [$^{32}$P] and incubated without or with an increasing concentration of nuclear proteins extracted from the liver of an adult male rat. Binding reactions included poly dI:dC (2 ug/10ul reaction) as a source of non-specific DNA sequences. The labeled DNA fragments are separated electrophoretically and visualized by autoradiography.

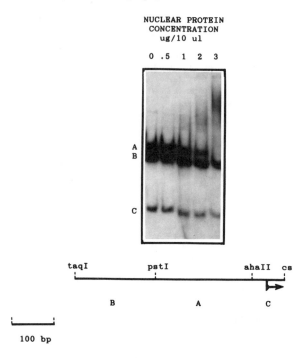

**Figure 2: Gel Shift Analysis of the Hss-1 Region.** A 487 bp insert was isolated from the pS14EXOT1-8cc plasmid. The insert spans the Hss-1 site and extends from -462 to +25 bp. The insert was digested with PstI and AhaII to generate 3 fragments which were end labeled with [$^{32}$P]. The labeled DNA's were incubated with nuclear proteins extracted from rat liver as described by Graves et al. (1986). Nuclear protein binding reactions included poly dI:dC at 2ug/10 ul and were carried out at 30°C. for 15 mins. DNA fragments were separated in 6% acrylamide:bisacrylamide (29:1) gels with 0.45M Tris-borate, pH 8.3 and 1.25 mM EDTA as buffer. After electrophoresis, gels were dried and exposed to x-ray film.

A major fraction of the "A" fragment displayed a shift in mobility with increasing concentration of rat liver nuclear proteins (Figure 2). The pattern of migration of the "A" fragment-protein(s) complex suggested multiple DNA-protein interactions. The lack of significant retardation of the "B" and "C" fragments indicate that binding of the rat liver nuclear proteins to the "A" fragment was sequence specific.

The analysis was extended to determine whether the sequence specific binding proteins in rat liver were tissue-specific. No specific retardation of the "A" fragment was detected when proteins from spleen. kidney or brain were added. The absence of DNA binding proteins which interact specifically with the "A" fragment in tissues which do not transcribe the S14 gene suggests that tissue-specific DNA binding proteins interact with sequences upstream from the S14 gene. These proteins may function not only in maintenance of the structure of Hss-1, but in the tissue-specific regulation of S14 gene transcription. Additional studies are required to establish the specific sequences recognized by these proteins and their role in S14 gene transcription and chromatin structure

## Developmental Regulation of S14 Gene Expression

In the rat, many hepatic enzymes involved in lipid and carbohydrate metabolism increase at weaning (Henning, 1978; 1981; Hemon, 1968; Vernon and Walker, 1968; Cook and Spence, 1973). This represents an adaptive response of the liver to a change in nutrient composition from a high fat milk diet to low-fat rat chow. The mRNA coding for the S14 protein showed a $\geq$100-fold increase during the period from 15 to 30 days post-partum (Jump and Oppenheimer, 1985). Recent studies showed that activation of S14 gene transcription was reported to account for the major increase in rat liver mRNAs$_{14}$ during post-natal development (Jump et al., 1988).

Prior to 15 days of age, the hepatic S14 gene was transcriptionally inactive and only Hss-2, 4 and 5 DNase I hypersensitive sites were present. Between 18 and 22 days of age, animals were weaned and hepatic S14 gene transcription and mRNAs$_{14}$ levels increased $\geq$ 40-fold and $\geq$ 100-fold, respectively. Accompanying this major increase in S14 gene transcription was the induction of the Hss-1 and Hss-3 DNase I hypersensitive sites. The ontogeny of S14 gene expression was associated with sequential changes in the chromatin organization of the S14 genomic domain. The fact that highly specific changes in chromatin structure coincide with gene activation suggest the gene exists in a poised state prior to activation requiring the presence of tissue-specific factors which interact near the S14 promoter at the Hss-1 site and upstream at the Hss-3 site to activate S14 gene transcription (Jump et al., 1988).

Finding that activation of S14 gene transcription occurred at the time of weaning suggested dietary composition may contribute to the expression of hepatic mRNAs$_{14}$ during post-natal development. Perez-Castillo et al., (1987) reported that weaning animals prematurely onto diets containing no-fat, but high-sucrose induced a rapid increase in hepatic mRNAs$_{14}$. In contrast, preliminary studies in our laboratory showed that weaning animals onto a high-fat diet (ICN: high fat diet) repressed the normal post-natal rise in hepatic mRNAs$_{14}$. These studies suggest the presence of dietary fat inhibits the normal development of S14 gene transcription. High dietary fat suppresses hepatic lipogenesis (Clark, 1986) and may influence the response of the liver to specific hormones. The recent study by Issad et al. (1987) showing that suckling rats displayed insulin resistance suggests the diet induced increase in S14 gene expression may be regulated by increasing the response of the liver to a specific hormone like insulin.

## T$_3$ Regulation of S14 Gene Expression

Thyroid hormone has a major effect on the tissue abundance of hepatic mRNAs$_{14}$ in the adult rat. T$_3$ administration to hypothyroid rats induced both the mRNAs$_{14}$ and its nuclear precursor within minutes of hormone administration suggesting $T_3$ acts at the transcriptional level (Jump et al., 1984; Narayan et al., 1984). However, other investigators reported high S14 transcriptional rates in hypothyroid rat liver with only marginal effects of $T_3$ on hepatic S14 transcriptional activity (Narayan and Towle, 1984; Tao and Towle, 1986). Instead of inducing S14 gene transcription, $T_3$ was suggested to increase hepatic $mRNAs_{14}$ by activating post-transcriptional mechanisms.

In contrast to these previous results, we recently reported that $T_3$ had a major effect on the regulation of S14 gene transcription (Jump, 1989). The relative level of $mRNAs_{14}$ abundance in the hypothyroid, euthyroid and hyperthyroid rat liver was found to be $0.22 \pm 0.05$, $1.0 \pm 0.07$ and $5.0 \pm 0.25$, respectively. The relative level of in vitro run-on activity for the S14 gene was $5.3 \pm 0.9$, $33.4 \pm 2.5$ and $43.8 \pm 2.0$ ppm in hypo-, eu- and hyperthyroid rat liver nuclei,respectively (Jump, 1989). Overall, $T_3$ induced a 8.3-fold increase in S14 gene transcription and a 23-fold increase in $mRNAs_{14}$.

In the transition from the hypothyroid to the euthyroid state, hepatic $mRNAs_{14}$ increased 4.5-fold which correlated with the 6-fold increase in S14 run-on activity. However, in the euthyroid to hyperthyroid transition, $mRNAs_{14}$ increased 5-fold while S14 run-on activity increased only 31%. Although the overall 8.3-fold increase in S14 transcriptional activity accounted for a major fraction of the 23-fold difference in hepatic $mRNAs_{14}$ in the hypo- to hyperthyroid transition, transcription alone could not account for the rise in $mRNAs_{14}$. These results show that $T_3$ induced both transcriptional and post-transcriptional mechanisms to augment $mRNAs_{14}$. Similar findings have been reported for the effects of $T_3$ on the induction of rat pituitary growth hormone (Diamond and Goodman, 1985) and hepatic malic enzyme (Dozin, et al., 1986) and HMG-CoA reductase (Simonet and Ness, 1988).

The significant difference in S14 transcriptional activity in the hypo- and hyperthyroid states led us to examine the kinetics of $T_3$ effects on S14 gene transcription. Following administration of a receptor-saturating dose of $T_3$ (200 ug/100 g.b.w.) to hypothyroid rats, a 5 min. lag period preceded a significant rise in both hepatic $mRNAs_{14}$ and S14 gene transcription. S14 gene transcription was induced 5-fold within 15 mins. and reached 70% of the maximal 9-fold induction within 2 hours. Hepatic $mRNAs_{14}$ levels increased in a linear fashion from 5 mins to 4 hours and was induced 24-fold above basal levels after 24 hours of $T_3$ treatment. The rapid induction of S14 gene transcription suggested $T_3$ acted directly at the chromatin level, presumably through the chromatin-associated $T_3$ receptor.

The disparity between the results reported from this lab (Jump, 1989) and the previous study by Narayan and Towle (1986) was attributed to the cDNA probes used in the hybridization analysis of S14 transcriptional activity. Either the DNA sequence characteristics of the pS14-C2 plasmid and/or the purity of plasmid DNA preparations used in the earlier studies contributed to the erroneous measurement of high transcriptional activity (Jump, 1989).

## $T_3$ Effects on S14 Chromatin Structure

Studies on the developmental and tissue-specific regulation of S14 chromatin structure suggest two sites located upstream from the S14 gene may be important in the regulation of S14 gene transcription. One site was adjacent to the S14 cap site and corresponds to the Hss-1 DNase I hypersensitive site, while the other site was located between -2.8 and -3.3 kb upstream from the end of gene and corresponds to the Hss-3 DNase I hypersensitive site. In order to determine whether these sites are directly influenced by $T_3$, effects on chromatin structure were examined in both thyroidal steady states and in the kinetic analysis.

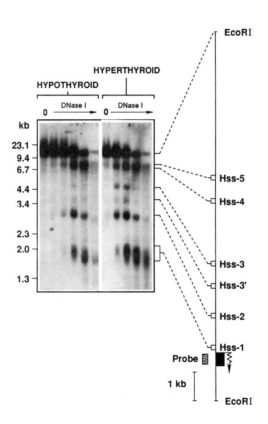

**Figure 3:** **Thyroidal Effects on DNase I Hypersensitive Sites Upstream from the S14 Gene.** Rats were made hypothyroid by maintenance on methimazole (0.25% in drinking water) for 4 weeks. Euthyroid rats were made hyperthyroid by administering $T_3$ i.p. (15 ug/100 g.b.w.) for 5 days. The isolation of nuclei and characterization of DNase I hypersensitive sites has been described previously (Jump et al., 1989).

Although hepatic S14 gene transcription was nearly 9-fold higher in hyperthyroid animals than in hypothyroid animals, no consistent difference in chromatin structure was detected at the Hss-1, 2, 4 or 5 DNase I hypersensitive sites flanking the 5'end of the S14 gene. However, the DNase I sensitivity of Hss-3' (-2.1 kb) and Hss-3 (-3.0 kb) hypersensitive sites was significantly increased in hepatic nuclei obtained from hyperthyroid animals than nuclei obtained from hypothyroid animals. The enhanced DNase I sensitivity of these structures correlated with the elevated transcriptional activity of the S14 gene measured in hyperthyroid liver nuclei.

In order to determine whether the Hss-3 and Hss-3' DNase I hypersensitive sites were rapidly induced by $T_3$, S14 chromatin structure was examined in hypothyroid animals receiving either vehicle or $T_3$ for 5, 15, and 120 mins. While the Hss-3 site was induced within minutes of hormone administration, the Hss-3' site remained unchanged. The DNase I sensitivity of the Hss-3 site was induced 2-fold within 5 mins and 4-fold within 2 hours of $T_3$ administration to hypothyroid rats. The induction of Hss-3 within 5 mins. of $T_3$ administration preceded the $T_3$-mediated increase in S14 gene transcription and suggested modification of chromatin structure at this site may be an antecedent event in the $T_3$ regulation of S14 gene expression (Jump, 1989). Thus, $T_3$ may initiate a change in S14 gene transcription by inducing a site specific modification of chromatin upstream from the S14 gene.

Although others reported on $T_3$ mediated changes in chromatin structure of the rat pituitary growth hormone gene (Nyborg and Spindler, 1986) and hepatic malic enzyme (Usala et al., 1988), these changes were measured at a time when $T_3$ induced maximal changes in gene transcription. The rapid $T_3$-mediated induction of the Hss-3 DNase I hypersensitive site located between -2.8 and -3.3 kb upstream from the S14 cap site represents the earliest effect of $T_3$ on nuclear structure noted to date. Moreover, the change in structure is highly specific since other DNase I hypersensitive sites within 11 kb of DNA flanking the S14 gene were not affected by acute $T_3$ administration. The rapid induction of this structure suggests modification of DNA-protein interaction at this locus is mediated by $T_3$ interaction with its chromatin-associated receptor.

**Nutritional Regulation of S14 Gene Expression**

The abundance of hepatic $mRNAs_{14}$ is also regulated by nutritional status (Jump et al., 1984; Mariash et al., 1986). In animals starved for 48 hours, hepatic $mRNAs_{14}$ levels were reduced to 10% of the level found in chow fed animals (Table 2). The reduction in $mRNAs_{14}$ was due to a 82% reduction in S14 transcriptional activity. Preliminary studies show that within 4 hours of administration of a sucrose gavage (1.0 ml of 60 % sucrose/100 g.b.w.) to 48 hour starved euthyroid rats, both hepatic S14 gene activity and $mRNAs_{14}$ abundance were induced to levels comparable to that found in chow-fed control animals. The induction of hepatic S14 transcriptional activity and $mRNAs_{14}$ abundance may be due to a rapid change in the plasma insulin:glucagon ratio induced by acute elevation in plasma glucose. These studies show that, in addition to $T_3$, other physiological stimuli can induce rapid changes in S14 gene transcription.

A comparison of hepatic S14 chromatin structure of 48 hr starved rats and chow fed animals is illustrated in Figure 4. Starvation induced a significant reduction of DNase I sensitivity of the Hss-1, 3 and 3'sites while having minimal effects on the structure of the Hss-2, 4 and 5 sites. The loss of Hss-1 and Hss-3 sites correlated with the 82% reduction in S14 gene transcription suggesting that maintenance of these structures was dependent on the nutritional status of the animal.

Table 2: Hepatic S14 Gene Expression in Chow-Fed and Starved Rats.

| Physiological State[a] | Relative mRNAs$_{14}$ Level[b] | S14 Transcriptional Activity (ppm) |
|---|---|---|
| Chow-Fed | 0.9 ± 0.05 (5)[c] | 33.4 ± 2.5 (5) |
| Starved (48 hrs.) | 0.1 ± 0.01 (3) | 4.5 ± 2.0 (3) |

[a]Male euthyroid rats (250-300 g.) were fed rat chow ad libitum or starved for 48 hours prior to the experiment.
[b]Euthyroid standard = 1.0 or approximately 1200 copies of mRNAs$_{14}$/cell.
[c]The value in parenthesis represents the number of animals. Data is presented as mean ± S.D.

## Summary

Both the chromatin structure and the transcription of the rat liver S14 gene are under complex control involving tissue-specific, developmental, hormonal and nutritional factors. Figure 5 summarizes results on the transcriptional regulation and chromatin structure of the rat liver S14 gene. The presence of 4 major and 2 minor DNase I hypersensitive sites flanking the 5'end of the gene correlates with the transcriptional activity of the S14 gene. The absence of the Hss-1, 2, 3' and 3 DNase I hypersensitive sites from tissues failing to express the S14 gene implicate these structures in the tissue-specific regulation of the gene. The induction of two tissue-specific sites, i.e., Hss-1 and Hss-3, during post-natal development and the deinduction of these same sites during starvation of the adult rat suggests nutritional status plays a major role in both maintenance of S14 chromatin structure and S14 gene activity.

Preliminary studies using streptozotocin-induced diabetic rats show diminished levels of hepatic mRNAs$_{14}$ and gene transcription and loss of the Hss-1 DNase I hypersensitive site. Kinlaw et al., (1987) reported inhibition of S14 gene transcription by glucagon. While elevated tissue cAMP levels may be inhibitory to S14 gene expression the presence of insulin-induced second messengers may be stimulatory to S14 gene transcription. Roesler et al., (1988) recently reviewed the evidence for cis-linked cAMP-responsive elements (CRE) associated with genes subject to positive regulation by cAMP-dependent A-kinase. However, elements involved in the cAMP-dependent inhibition of gene transcription or in the insulin-dependent regulation of gene transcription have not been identified. These studies suggest that second messengers generated from the plasma membrane function in concert with T$_3$ and tissue-specific factors to regulate S14 gene transcription.

The specificity and the kinetics of induction of the Hss-3 site located between -2.8 to -3.3 kb upstream from the S14 gene suggest T$_3$ acts through its chromatin-associated receptor to locally modify S14 chromatin structure. Modification of chromatin structure in the vicinity of the Hss-3 may play a causal role in the T$_3$-mediated induction of S14 gene transcription (Jump, 1989). Several studies describe the presence of enhancer elements within inducible DNase I hypersensitive sites flanking the 5'end of active genes (Parslow and Granner, 1982,1983; Zaret and

Yamamoto, 1984; Jantzen et al., 1987; Hecht et al., 1988). In most cases, the inducible sites correspond to the location of steroid hormone binding sites within chromatin, i.e., hormone responsive elements (HRE). Binding of the ligand-activated receptor to chromatin induces a local change in DNA-protein interaction around the binding site and alters the transcription rate of cis-linked genes (Cordingley et al., 1987). The linkage between these two events, however, is unclear and will require further study. Whether the $T_3$ induction of the Hss-3 DNase I hypersensitive site harbors a thyroid hormone-responsive element(s) and is involved in activation of transcription of the cis-linked S14 gene will also require additional studies.

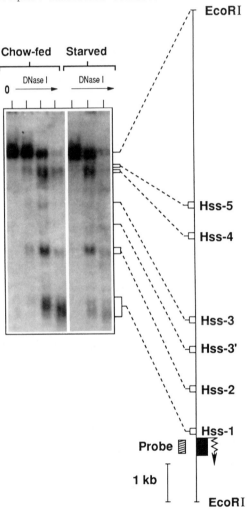

**Figure 4: Effects of Starvation on S14 Chromatin Structure.** Chow-fed male euthyroid rats were maintained on rat chow ad libitum. Male euthyroid rats were starved for 48 hours. The isolation of nuclei and characterization of DNase I hypersensitive sites has been described previously (Jump et al., 1988).

Finally, the complex regulation of S14 gene transcription by tissue-specific, developmental, hormonal and nutritional factors may require multiple functional elements. Although additional studies are required, the chromatin studies described above provide preliminary evidence for the location of two such functional elements. Tissue-specific and nutritional factors regulated the Hss-1 site, while T₃ regulates the Hss-3 site. Studies on steroid hormone regulation of murine mammary tumor virus and tryptophan oxygenase gene suggest that factors in addition to the steroid hormone receptor are required for optimal function of hormone-responsive elements cis-linked to active genes (Butti and Kuhnel, 1986; Miksicek et al., 1987; Danesch et al., 1987; Schule et al., 1988). Defining the nature of the interaction between hormone receptors and other chromatin elements involved in gene transcription remains a central issue in molecular endocrinology.

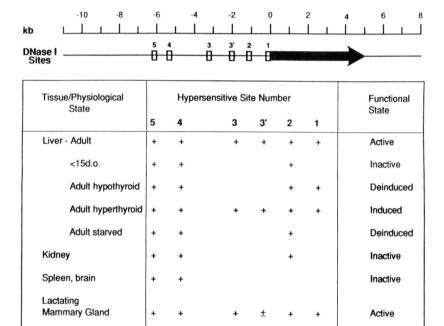

| Tissue/Physiological State | Hypersensitive Site Number | | | | | | Functional State |
|---|---|---|---|---|---|---|---|
| | 5 | 4 | 3 | 3' | 2 | 1 | |
| Liver - Adult | + | + | + | + | + | + | Active |
| <15d.o. | + | + | | | + | | Inactive |
| Adult hypothyroid | + | + | | | + | + | Deinduced |
| Adult hyperthyroid | + | + | + | + | + | + | Induced |
| Adult starved | + | + | | | + | | Deinduced |
| Kidney | + | + | | | + | | Inactive |
| Spleen, brain | + | + | | | | | Inactive |
| Lactating Mammary Gland | + | + | + | ± | + | + | Active |

Figure 5: Summary of the Transcriptional Activity and Chromatin Structure at the Rat Liver S14 Gene Locus. The transcription unit is depicted as a black horizontal arrow. The location of the 6 DNase I hypersensitive sites are identified by their respective number. The scale above the gene map indicates the area analyzed (Jump et al., 1987; Jump et al, 1988; Jump, 1989). The tissue or physiological state, the presence and absence of the DNase I hypersensitive site and functional state of the gene are noted.

## Acknowledgments

This work was supported by the National Institute of General Medical Sciences (GM36851). I would like to thank Andrew Veit and Vivian Santiago for excellent technical assistance and Peter Carrington for the preparation of the figures.

## References

Alpers, D.H., Strauss, A.W., Ockner, R.K., Bass, N.M., Gordon, J.I. (1984) Proc. Natl. Acad. Sci. U.S.A. 81: 313-317.

Brindley, D.N. (1985) in: Biochemistry of Lipids and Membranes. (Vance, D.E. and Vance, J.E., eds), pp. 213-241, The Benjamin/Cummings Publishing Co., Inc. Menlo Park, Ca.

Butti, E. and Kuhnel, B. (1986) J. Mol. Biol. 190: 379-391.

Clarke, S.D., (1986) in: Dietary Fat and Cancer (Ip, C., Birt. D.F., Rogers, A.E. and Mettlin, C. eds) pp 531-553, Alan R. Liss, Inc., NY.

Clayton, D.F. and Darnell, J.E., (1983) Mol. Cell. Biol. 3: 1552-1561.

Cook, H.W. and Spence, M.W., (1973) J. Biol. Chem. 248: 1793-1796.

Cook, H.W. (1985) in: Biochemistry of Lipids and Membranes. (Vance, D.E. and Vance, J.E. eds.), pp. 181-212, The Benjamin/Cummings Publishing Co., Inc. Menlo Park, CA.

Cordingley, M., Riegel, A. and Hager, G. (1987) Cell 48: 261-270.

Danesch, U., Gloss, B., Schutz, G., Schule, R. and Renkawitz, R. (1987) EMBO J. 6: 625-630.

Diamond, D.J. and Goodman, H.M. (1985) J. Mol. Biol. 181: 41-62.

Dillman, W.H., Surks, M.I., and Oppenheimer, J.H. (1974) Endocrinology 96: 492-498.

Dozin, B., Magnuson, M.A. and Nikodem, V.M. (1986) J. Biol. Chem. 261: 10290-10292.

Elgin, S.C.R. (1981) Cell 27: 413-415.

Evans, R.M. (1988) Science 240: 889-895.

Flug, F., Copp, R.P., Casanova, J., Horowitz, Z.D., Janocko, L., Plotnick, M. and Samuels, H.H. (1987) J. Biol. Chem. 262: 6373-6382.

Freake, H. and Oppenheimer, J.H. (1987) Proc. Natl. Acad. Sci. U.S.A. 84: 3040-3074.

Fried, M. and Crothers, D.M. (1981) Nucleic Acids Res. 9: 6505-6525.

Gardner, G., Durand, G., and Pascal, G. (1983) Lipids 18: 223-225.

Garner, M.M. and Revzin, A. (1981) Nucleic Acids Res. 9: 3047-3060.

Glass, C.K., Franco, R., Weinberger, C., Albert, V.R., Evans, R.M., Rosenfeld, M.G. (1987) Nature 329: 738-741.

Godowski, P., Picard, D. and Yamamoto, K. (1988) Science 241: 812-816.

Goodridge, A. (1985) in: Biochemistry of Lipids and Membranes. (Vance, D.E. and Vance, J.E. eds.), pp. 143-180, The Benjamin/Cummings Publishing Co., Inc., Menlo Park, CA.

Graves, B.J., Johnson, P.F. and McKnight, S.L. (1986) Cell 44: 565-576.

Gross, D. and Garrard, W. (1988) Ann. Rev. Biochem. 57: 159-197.

Hecht, A., Berkenstam, A., Stromstedt, P.-E., Gustafsson, J.-A. and Sippel, A. E. (1988) EMBO J. 7: 2063-2073.

Hemon, P. (1968) Biochim. Biophys. Acta 151: 681-683.

Henning, S.J. (1978) Am. J. Physiol. 235: E451-E456.

Henning, S.J. (1981) Am. J. Physiol. 241: G199-G214.

Issad, T.. Coupe, C., Ferre, P., and Girard, J. (1987) Am. J. Physiol. 253: E143-E148.

Izumo, A., Nadal-Ginard, B. and Mahdavi, V. (1986) Science 231: 597-600.

Jantzen, H.M., Strahle, U., Gloss, B., Stewart, F., Schmid, W., Boshart, M., Miksicek, R. and Schutz, G. (1987) Cell 49: 29-38.

Jump, D.B., Narayan, P., Towle, H.C. and Oppenheimer, J.H. (1984) J. Biol. Chem. 259: 2789-2797.

Jump, D.B. and Oppenheimer, J.H. (1985) Endocrinology 117: 2259-2266.

Jump, D.B., Tao, T-Y, Towle, H.C. and Oppenheimer, J.H. (1986) Endocrinology 118: 1892-1896.

Jump, D.B., Wong, N.C.W., Oppenheimer, J.H. (1987) J. Biol. Chem. 262: 778-784.

Jump, D.B., Veit, A.M., Santiago, V., Lepar, J., Herberholz, L. (1988) J. Biol. Chem. 263: 7254-7260.

Jump, D.B. (1989) J. Biol. Chem. 264: In Press.

Kinlaw, W., Schwartz, H. and Oppenheimer, J. (1987) Clin. Res. 35: 398A.

Koenig, R.J., Brent, G.A., Warne, R.L., Larson, P.R. and Moore, D.D. (1987) Proc. Natl. Acad. Sci. U.S.A. 84: 5670-5674.

Liaw, C. and Towle, H.C. (1984) J. Biol. Chem. 259:7253-7260.

Mariash, C.N., Seelig, S., Schwartz, H.L. and Oppenheimer, J.H. (1986) J. Biol. Chem. 261: 9583-9586.

Miksicek, R., Borgmeyer, U. and Nowock, J. (1987) EMBO J. 6: 1355-1360.

Narayan, P. and Towle, H.C. (1985) Mol. Cell. Biol. 5: 2542-2646.

Narayan, P., Liaw, C.W., and Towle, H.C. (1984) Proc. Natl. Acad. Sci. U.S.A. 81: 4687-4691.

Nicols, D.G. and Lock, R.M. (1984) Physiol. Rev. 64: 1-64.

Nyborg, J. and Spindler, S.R. (1986) J. Biol. Chem. 261: 5685-5688.

Oppenheimer, J.H., Schwartz, H.L., Mariash, C.N., Kinlaw, W.B., Wong, N.C.W., and Freake, H.C., (1987) Endocrine Reviews 8: 288-307.

Parslow, T.G. and Granner, D.K. (1982) Nature 299: 449-451.

Parslow, T.G. and Granner, D.K. (1983) Nucleic Acids Res. 11: 4775-4792.

Perez-Castillo, A., Schwartz, H.L., Oppenheimer, J.H. (1987) Am. J. Physiol. 253: E536-E542.

Roesler, W., Vanderbark, G., and Hanson, R. (1988) J. Biol. Chem. 263: 9063-9033.

Sap, J., Munoz, A., Damm, K., Goldberg, Y., Ghysdael, J., Leutz, A., Beug, H., and Vennstrom, B. (1986) Nature 324: 635-640.

Schule, R., Muller, M., Otsuga-Murakami, H. and Rendawitz, R. (1988) Nature 332: 87-90.

Schutz, G., Schmid, W., Jantzen, M., Danesch, U., Gloss, B., Strahle, U., Becker, P., and Boshart, M. (1986) New York Acad. Sci. 478: 93-108.

Seelig, S., Liaw, C., Towle, H.C. and Oppenheimer, J.H. (1981) Proc. Natl. Acad. Sci. U.S.A. 78: 4733-4737.

Seelig, S., Jump, D.B., Towle, H.C., Liaw, C., Mariash, C.N., Schwartz, H.L., and Oppenheimer, J.H. (1982) Endocrinology 110: 671-673.

Shupnik, M., Chin, W.W., Habener, J.F., and Ridgeway, E.C. (1985) J. Biol. Chem. 260: 2900-2903.

Simonet, W.S. and Ness, G.C. (1988) J. Biol. Chem. 263: 12448-12453.

Surks, M.I., Koerner, D.H., and Oppenheimer, J.H. (1975) J. Clin. Invest. 55: 50-60.

Vernon, R.G. and Walker, D.G. (1986) Biochem. J. 106: 321-329.

Tao, T-Y. and Towle, H.C. (1986) Ann. New York Acad. Sci. 478: 20-30.

Usala, S.J., Young, W.S., III., Morioka, H. and Nikodem, V.M. (1988) Mole. Endo. 2: 619-626.

Waterman, M.L., Alders, S., Nelson, C., Greene, G.L., Evans, R.M., and Rosenfeld, M.G. (1988) Mole. Endo. 2: 14-21.

Weinberger, C., Thompson, C.C., Ong, E.S., Lebo, R., Groul, D., and Evans., R.M. (1986) Nature 324: 641-646.

Weintraub, H. and Groudine, M. (1976) Science 193: 848-856.

Wight, P.A., Crew, M.D., and Spindler, S.R., (1987) J. Biol. Chem. 261: 5685-5688.

Wight. P.A., Crew, M.D., and Spindler, S.R. (1988) Mole. Endo. 2: 536-542.

Weisbrod. S. (1982) Nature 297: 289-295.

Wolff, E.C. and Wolff, J. (1969) The Thyroid Gland (Pitt-Rivers, R. and Trotter, W.R., eds.) Vol. 1, pp. 237-282.

Wu, C. (1980) Nature 286: 854-860.

Yaffe, B.M. and Samuels, H.H. (1984) J. Biol. Chem. 259: 6284-6291.

Yamamoto, K.R. (1985) Annu. Rev. Genetics 19: 209-252.

Ye, Z-S., Forman, B.M., Aranda, A., Pascual, A., Park, H-Y., Casanova, J. and Samuels, H.H. (1988) J. Biol. Chem. 263: 7821-7829.

Zaret, K.S. and Yamamoto, K.R. (1984) Cell 38: 29-38.

DISCUSSION OF THE PAPER PRESENTED BY D. JUMP

TSAI: Is there any TRE in the #3 hypersensitive site? What do you mean by post-transcriptional modification of #3 site?

JUMP: We have preliminary DNA sequence data extending from the S14 cap site to -3.5 kb upstream. We find 3 sites showing 60-80% homology to the thyroid hormone regulatory element described by growth hormone (Glass et al., [1987] Nature 329:738). These sites are located at -2.15, -2.9 and -3.2 kb upstream from the S14 gene. The Hss-3' site maps at the -2.1 kb and the $T_3$-inducible Hss-3 site maps at the -3.2 kb region. We are curretnly determining whether $T_3$ receptors bind specifically to DNA sequences at these locations.

While the Hss-3 site at -3.2 kb is induced within minutes by $T_3$, the Hss-3' site at -2.1 kb is induced by $T_3$ more slowly. This structure is most evident in nuclei derived from hyperthyroid animals. The induction of the Hss-3' occurs after the S14 gene is transcriptionally activated by $T_3$. We refer to this as a post-transcriptional modification of chromatin structure. The kinetic relationship between the activation of the S14 gene transcription and the induction of the Hss-3' site suggest the modification of chromatin structure in the vicinity of the Hss-3' site may not be crucial to S14 gene transcription. The significance of the change in Hss-3' structure to S14 gene transcription is not clear.

HAQUE: Is S14 mRNA related to $T_3$? Does reverse $T_3$ have an effect on hepatic $RNAs_{14}$? Is there any such marker for reverse $T_3$?

JUMP: The relationship between $T_3$ and $mRNAs_{14}$ is that $T_3$ induces a rapid increase in the transcription of the S14 gene and also activates post-transcriptional processes. These two events account for the $T_3$ effect on the hepatic concentration of $mRNAs_{14}$. We do not think the S14 protein is involved in $T_3$ metabolism. Reverse $T_3$ has no effect on hepatic mRNAs$_{14}$ abundance. We are not aware of any speciic protein or mRNA sequence affected by reverse $T_3$. Reverse $T_3$, binds with very low affinity to the nuclear receptor for $T_3$.

CASTRILLO: Is this the typical resolution with this method of analysis of DNase I hypersensitive sites?

JUMP: This is the typical resolution of a DNase I hypersensitive site analysis using the methods developed in our laboratory. We estimate the size of the DNase I hypersensitive sites to approximate about one nucleosome repeat length, i.e., 150-200 base pairs of DNA. Other techniques are available to examine DNase I hypersensitive site fine structure such as the in vivo DNase I footprinting technique based on the genomic sequencing method developed by Church and Gilbert.

THOMPSON: Have you been able to carry out any nucleosome phasing experiments:

JUMP: We have not carried out specific experiments which address nucleosome phasing. However, we do expect to see phasing between the hypersensitive sites.

THOMPSON: At the Gordon Conference on Hormone Action last August, G. Mueller said he had data for histone-steroid hormone receptor interaction. Does the thyroid hormone do so?

JUMP: We have no direct evidence for an interaction between $T_3$ receptors and nucleosomal histones or histone H1. Studies from Norman Eberhardt and colleagues suggest that nucleosomal histones may stabilize $T_3$ receptors (Eberhardt et al., [1979] Proc. Natl. Acad. Sci. USA 76: 5005.

DISCUSSANTS: M. TSAI, D. JUMP, N. HAQUE, J.L. CASTRILLO AND E.B. THOMPSON.

# Hormonal and Developmental Regulation of Xenopus Egg Protein Genes

J.R. Tata, H. Lerivray, J. Marsh and S.C. Martin

Laboratory of Developmental Biochemistry, National Institute for Medical
Research, The Ridgeway, Mill Hill, London NW7 1AA, England.

## Introduction

### Multiple Sites of Formation of Egg Constituents

The developing oocyte alone is not capable of generating all the
different proteins, nucleic acids, lipids, metabolites, etc. that make
up the mature egg. Oogenesis is therefore a most appropriate example in
development of a precisely coordinated division of labor between
different tissues of an organism. The reader is referred to some
general reviews dealing with questions of oogenesis and egg development
(Browder, 1985; Metz and Monroy, 1985; Wallace, 1985).

### Components Generated within the Developing Oocyte

A large variety of macromolecules stored in the mature egg are
synthesized within the developing oocyte itself. Many of these
constitute the machinery for post-fertilization cell division and
protein synthesis, i.e. DNA and RNA polymerases, histones and non-
histone chromatin proteins, ribosomes, transcription and translation
factors, etc. Another group of macromolecules synthesised in situ
during oogenesis are messenger RNAs of great diversity, which are known
to include those coding for cytoskeletal and membrane proteins (Tata,
1986, 1988). Whatever the similarities or differences in the pattern of
stored mRNAs and proteins in eggs of different organisms, what is
important to note is that their formation in the oocyte is autonomous
and not regulated by hormones or other environment-linked factors.

### Extra-oocyte Synthesis of Egg Components

There are three extra-oocyte tissues in which are synthesized
constituents destined for the egg, or hormones that control these
processes, namely the oviduct, liver and follicle cells. Two particular
features of the involvement of the liver and oviduct in participating in
egg development distinguish them from the 'autosynthetic' activity of
the oocyte itself. First, these tissues contribute a few specialized
proteins, lipids, etc., but in such large amounts that they may
constitute the bulk of the egg by mass (yolk, white proteins, jelly,
shell, etc.). Second, unlike the oocyte, their activity concerning egg

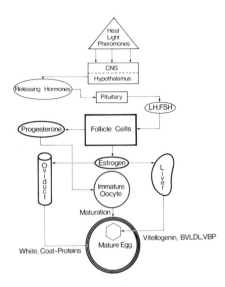

Figure 1. Coordination of hormone regulated functions participating in the process of egg development. CNS, central nervous system; LH, luteinising hormone; FSH, follicle stimulating hormone; BVLDL, low density lipoproteins; VBP, vitamin binding proteins.

development is entirely and obligatorily under hormonal control - a control essential for coordinating and synchronising the activities of different sites.

## Coordination of Activities of Multiple Sites involved in Oocyte Development: The Role of Hormones

It is essential that the activities of diverse sites engaged in the formation of egg components be tightly coordinated. This coordination is achieved by hormones. Figure 1 summarizes the participation of major hormones in oogenesis in vertebrates. The primary signal triggering off egg maturation is environmental which, in turn, is converted from electrical to chemical signals in the central nervous system (Tata, 1985). The hypothalamus secretes the decapeptide, gonadotropin releasing hormone (GnRH) in response to an external stimulus, which then activates the anterior pituitary to release gonadotropins (LH and FSH). The latter stimulate follicle cells to secrete estrogen and progesterone with the consequent stimulation of egg protein synthesis in liver and oviduct (Tata, 1986). In some species, such as the Xenopus, progesterone acts directly on the oocyte via the cyclic AMP-$Ca^{++}$ pathway to initiate meiotic division.

The hormonal coordination of egg maturation is highly conserved in all vertebrates. An almost similar neuroendocrine regulation of oogenesis operates in invertebrates with a different set of hormones. It is therefore interesting to consider the hormonal regulation in an evolutionary context. In view of the importance of egg development, it is obvious that the hormone receptors in tissues involved in oogenesis would also be highly conserved during evolution. As estrogen and progesterone have remained chemically unchanged during evolution, the hormone-binding and functional domains of the receptor in oviduct and liver also would not have diverged.


## The Liver and Vitellogenesis

The liver contributes a substantial proportion of the total mass of the egg in the form of yolk proteins, low density lipoproteins, vitamin-carrier proteins, as well as cholesterol and phospholipids. These are synthesized in the liver under obligatory estrogen control and secreted into the blood from which they are taken up by the oocyte via specific receptors and endopinocytosis (Wallace, 1985; Tata, 1988). The regulation in liver of genes encoding vitellogenin, the precursor of yolk proteins, is a popular model system for studying steroid hormonal control of gene expression. Among its major advantages is that the quiescent vitellogenin genes in male liver can be activated by the hormone exactly as in the normal female, thus allowing a better analysis of the first stages of de novo gene activation. Furthermore, the physiological process of vitellogenin synthesis and secretion in vivo can be fully and reversibly reproduced in primary cultures of hepatocytes in the absence of any cellular proliferation following the addition and withdrawal of estrogen (Tata et al., 1985; Tata, 1987).


## The Xenopus Vitellogenin Gene Family

The four actively expressed vitellogenin genes of Xenopus laevis fall into two groups and are termed A1, B1, A2 and B2 (Wahli et al., 1979). There is 80% homology of coding sequence between groups A and B, and 95% between two member genes of each group. Each is made up of 34 similar exons but the size of the gene varies from 16 to 21 kbp of DNA because of large variations in intron sizes. The A1, A2 and B1 genes are linked with 15 kbp of DNA between A1 and B1 genes (Wahli et al., 1979,1982; Schubiger and Wahli, 1986). This arrangement within a gene family raises the question of whether or not the individual gene members are expressed coordinately and to the same extent when under hormonal control. Sequencing of the 5' upstream flanking regions (Walker et al., 1984; Germond et al., 1984) has revealed both similarities and differences within the gene family which may provide clues to the possible regulatory regions with which the estrogen receptor complex may interact.


## Hormonal Induction of Vitellogenin Genes

The complete absence of any vitellogenin-like material or its mRNA throughout its life in the adult male vertebrate liver or blood makes it easier to establish the early stages of hormonal induction of yolk proteins in the male than in the female. From a careful hybridization analysis, Baker and Shapiro (1977) were the first to show in whole animals that no vitellogenin mRNA could be detected in the male Xenopus liver and that the differences in primary and secondary responses seen in circulating vitellogenin were preceded by parallel changes in vitellogenin mRNA in male Xenopus liver. It was also shown that these mRNA levels exceeded those of albumin and that the kinetics and extent

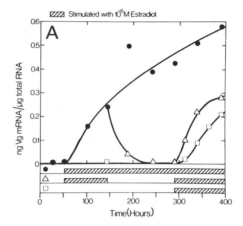

Figure 2. Kinetics of vitellogenin mRNA accumulation during primary and secondary induction in the same batch of cultured cells. Hepatocyte cultures were prepared from 8 hormonally unstimulated male Xenopus livers and stimulated with 10-$^6$M estradiol in culture for the periods indicated by the hatched bars below the graph. Total cellular RNA was extracted from the cells after 1-16 days in culture and the vitellogenin mRNA content was measured by disc hybridization to plasmid E7 carrying Xenopus vitellogenin cDNA insert. (●), Cells exposed to estradiol continuously after 2 days in culture; (Δ), after 4 days of primary stimulation, the cultured cells were washed and withdrawn from hormone stimulation for 6 days, after which 10-$^6$M estradiol was added to the medium (secondary induction); (□), estrogen was first added after 12 days in culture. (Data adapted from Searle and Tata, 1981.)

of accumulation of vitellogen mRNA during secondary induction in male and female liver were very similar. However, studies in whole animals have serious drawbacks, if one is to address the central issue of the mechanisms underlying the role of hormone-receptor interactions with the induced gene during early stages of the onset of gene activation. For these reasons, the authors' laboratory has devoted much attention to studying the regulation of vitellogenin gene expression in primary cell cultures.

## Regulation of Expression of Vitellogenin Genes in Primary Cultures of Male Xenopus Hepatocytes

Among the many advantages of cell culture over whole animals in studying hormonal regulation of gene expression are: a) the ability to control hormone concentration accurately; b) better analysis of the early events associated with hormone-receptor interaction with regulatory elements of the gene; c) rapid reversibility of induction by removal of the hormone thus enabling the study of the de-induction process; d) the analysis of single cell types in heterogeneous tissues pooled from many animals, thus reducing variability and interference from non-competent cells.

It has often been observed that freshly prepared primary cell cultures respond poorly to various stimuli, including hormones, nutrients, drugs, etc. (Wolffe and Tata, 1984). The recognition of this "culture shock" phenomenon made it possible to reproduce in primary cell cultures the de novo activation of vitellogenin genes and to sustain the accumulation of vitellogenin mRNA at high rates for several days in the continued presence of estrogen. By allowing cells to recover from culture shock the physiological characteristic of primary and secondary inductions can be reversibly reproduced in primary culture (Searle and Tata, 1981; Wolffe and Tata, 1983).

The steady-state levels of vitellogenin mRNA induced de novo by a single administration of estrogen in cultures of Xenopus hepatocytes, as shown in Figure 2 (Searle and Tata, 1981; Wolffe and Tata, 1983; Wolffe, 1984), was initially due to transcription under conditions in which this mRNA is highly stable. When the stability of vitellogenin mRNA was directly measured in the continuous presence or after withdrawal of estrogen from cultures, it was found that the presence of the hormone stabilized vitellogenin mRNA. Thus, whereas the t of vitellogenin mRNA in the continuous presence of estrogen was >48 hr, the removal of the hormone caused this value to fall to <16 hr. Similar results had been described by Brock and Shapiro (1983). What is of particular importance is that the stabilization by estrogen was specific for the induced mRNA, and indeed the hormone has the opposite effect on the constitutively expressed albumin mRNA. This latter effect may in part explain the deinduction of albumin that accompanies induction of vitellogenin synthesis (Tata and Smith, 1979; Farmer et al., 1978). The mechanisms underlying the regulation of stability of hormone-induced mRNAs remain unknown.

## Differential Activation of Individual Xenopus Vitellogenin Genes

Analysis of RNA extracted at early periods after the addition of estradiol to primary cultures of male Xenopus hepatocytes (Ng et al., 1984), as depicted in Figure 3, showed quite clearly that the transcription and steady-state levels of individual vitellogenin mRNAs were not regulated coordinately or to the same extent. The rate and extent of transcription and accumulation of vitellogenin mRNA varied in the order of B1 > A1 > A2 $\cong$ B2 following addition of hormone to both adult male and female Xenopus hepatocytes.

## Estrogen Receptor and Vitellogenin Gene Activation

Adult male Xenopus liver has low levels of estrogen receptor, comprising only 200-500 molecules tightly bound to the nucleus per cell (Westley and Knowland, 1978; Hayward et al., 1980; Perlman et al., 1984). Treatment of naive male Xenopus with estrogen causes a 5-15 fold increase in high-affinity liver nuclear receptors to reach levels found in female liver (Westley and Knowland, 1979; Hayward et al., 1980; Perlman et al., 1984). This elevated level of receptor in male hepatocytes persists for several weeks so that it may also explain the more rapid and extensive response to the hormone during secondary induction, in addition to any long-lasting changes in the chromatin conformation of vitellogenin genes. Figure 4 clearly shows the close relationship between functional nuclear receptor and activation of the dormant genes by estrogen added to primary hepatocytes. Stimulation with the hormone caused the receptor level in naive male Xenopus cells to rise to those found in female cells, accompanied by similar enhancement of vitellogenin gene transcription to rates observed in female hepatocytes. Experiments with cycloheximide added at different

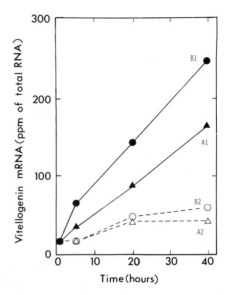

Figure 3. Accumulation of the four vitellogenin mRNAs measured separately in livers of stage 66 Xenopus froglets as a function of time of exposure to estradiol-17ß. Late metamorphic, or just postmetamorphic, larvae were kept in water containing $10^{-6}$M estradiol-17ß for 3 days, after which froglets at stage 66 were selected. Liver RNA was isolated at the times indicated, hybridized separately to HindIII excised fragments of vitellogenin cDNA corresponding to vitellogenin genes A1 (▲), A2 (△), B1 (●), and B2 (○), and the amount of each vitellogenin mRNA calculated. (From Ng et al., 1984.)

times of culture showed that the small amount of receptor residing in male liver nuclei at the start of the experiment accounted for the activation of gene transcription in the first 4 hr after which the increase in transcription required continuing protein synthesis for both processes. Recently, by carrying out hybridization analysis with cDNA to Xenopus estrogen receptor (Weiler et al., 1987), Shapiro's laboratory have shown that the upregulation of estrogen receptor by estrogen involves an activation of transcription of estrogen receptor gene in male Xenopus liver (Shapiro, unpublished data). In addition to changes in the number of receptor molecules, one has also to consider the more short-term modulation of receptor activity caused by reversible co-valent modifications upon hormone addition or withdrawal. The most likely modification is phosphorylation and de-phosphorylation, as has been demonstrated for estradiol receptor in the uterus (Migliaccio et al., 1986).

## Steroid Receptor and Regulatory Gene Sequences

The most likely explanation for the differential activation of the individual members of the vitellogenin gene family (Figure 3) may be different promoter strengths or variable intensities of interaction

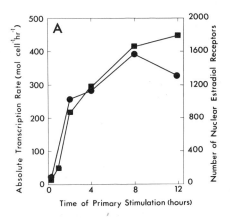

Figure 4. Correlation of nuclear estrogen receptor levels (●) with absolute rates of vitellogenin gene transcription (■) in male <u>Xenopus</u> hepatocytes as a function of time (hr) after the addition of estradiol. (Data adapted from Perlman et al., 1984.)

between estrogen receptor or other transcription factors and gene sequences bearing regulatory elements, termed estrogen response element (ERE). It is therefore most relevant that, while all four <u>Xenopus</u> vitellogenin genes have one or two ERE's as the imperfect palindromic sequence GGTCANNNTGACC between -310 and -375 bp in the 5' upstream region, the linked genes A1 and B1, which are more strongly expressed than the pair A2 and B2, have an additional element further upstream at -663 and -554 bp, respectively. Recently, the groups of Wahli and Ryffel have studied the estrogen-regulated expression in human MCF-7 breast cancer cells of a hybrid gene formed by fusion of the promoter regions of <u>Xenopus</u> vitellogenin genes B1 and A2 and the coding region of the bacterial CAT gene (Seiler-Tuyns et al., 1986; Klein-Hitpass et al., 1986). Transfection studies and deletion mapping showed that the 13 bp element at position -334 was essential for hormonal inducibility. It is worth noting that this element is also present in the 5' upstream region of other liver-specific estrogen-inducible genes, including the chicken vitellogenin gene Vtg II (Walker et al., 1984). The latter, which is more strongly expressed than the other two chicken vitellogenin genes, also has three ERE's at similar location 5' upstream from the transcription initiation site.

Considerable progress has been made with nuclease protection or foot-printing procedures for determining regions around steroid response elements that interact with the relevant receptor. Thus, DNA sequences located between -100 and -700 bp upstream from the transcription initiation site have been implicated as the site of regulation by many steroid hormone receptors of a variety of genes, such as ovalbumin, lysozyme, uteroglobin and MMTV (Eriksson and Gustafsson, 1983; Chambon et al., 1984; Renkawitz et al., 1984; Payvar et al., 1983; Dean et al., 1984). A consensus sequence located at -458 to -725 bp upstream in the 5' flank of the chicken vitellogenin gene II was found to be a site of interaction with estrogen receptor, as judged from DNase I protection assays (Jost et al., 1984). Similar core sequences have been detected in the 5' flank of all four Xenopus vitellogenin genes (Walker et al., 1984). Also, the chicken apoVLDL gene which is estrogen-regulated, but not induced de novo, in the liver has two similar sequences at around -300 bp upstream but not around -600 bp.

## Receptor-Gene Interaction as Studied by Cell Transfection

The technique of DNA transfection into cultured cells has contributed substantially to enhancing our understanding of structural requirements for receptor-DNA interactions. This is true for both delineating sequences in Xenopus vitellogenin genes which represent regulatory sites for estrogen receptor (Seiler-Tuyns et al., 1986), as well as generally for all genes regulated by steroid hormones (Green and Chambon, 1987). However, such cell transfection studies have mostly been carried out in heterologous cells or in transformed cell lines, in which some of the tissue-specific factors and other physiological constraints for regulation may be absent or modified. For this reason we undertook a comparison of the regulation by estrogen of the transcription of Xenopus vitellogenin promoter sequences linked to the reporter gene of bacterial chloramphenicol acetyl transferase (CAT) transfected into a Xenopus cell line (XTC2) and primary cultures of female Xenopus hepatocytes.

A major difficulty in undertaking such a comparison was the low efficiency of transfection of cells in primary culture with calcium phosphate- or dextran-based procedures, most frequently used for cell lines or tumor cells. In line with an approach also adopted by other workers recently (Tur-Kaspa et al., 1986; Chu et al., 1987), we decided to transfect DNA into Xenopus cell lines and hepatocyte cultures by the technique of electroporation. Field strength, composition of the buffer, the design and volume of electroporation, voltage, the state of DNA, time after transfection and the type of cells used, were all found to be critical factors in determining transfection efficiency based on CAT assays. Once optimum conditions were defined we used the Xenopus embryo-derived cell line XTC2 (presumably of epithelial origin) to compare the transcription of various vitellogen promoter-CAT constructs in the presence and absence of estrogen with that in primary cell cultures.

Using the same construct comprising the three ERE's 5' upstream of the vitellogenin B1 gene (-596 to +8) as used by Wahli's group (Seiler-Tuyns et al., 1986), transfection of XTC2 cells produced a high basal activity which prevented the detection of an induction with estrogen. The basal activity was attenuated by lowering the amount of DNA transfected, but estrogen failed to produce any stimulation (lanes 4 and 5, Figure 5). This result is comparable to the stimulation by estrogen of vit-CAT constructs transfected into the Xenopus kidney cell line B3.2 (Seiler-Tuyns et al., 1986). Since no vitellogenin mRNA could be detected in XTC2 cells treated with estrogen, it was possible that this cell line lacked functional receptor. It is therefore significant that upon co-transfection of this construct with a construct made by us incorporating a full-length cDNA to chicken estrogen receptor (Krust et al., 1986), kindly provided by Professor P. Chambon, Strasbourg, it was possible to obtain a response to estrogen (Figure 5, lanes 2 and 3).

In contrast to our results with XTC2 cells, primary hepatocyte cultures responded to estrogen without co-transfection with the ER cDNA of the same vitellogenin promoter construct (Figure 6, lanes 3 and 4). Figure

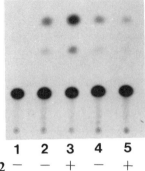

$$\begin{array}{ccccc} \textbf{1} & \textbf{2} & \textbf{3} & \textbf{4} & \textbf{5} \end{array}$$
$$E_2 \quad - \quad - \quad + \quad - \quad +$$

Figure 5. Co-transfection of <u>Xenopus</u> vitellogenin B1 gene-CAT construct and the chicken estrogen receptor (ER) expression construct into XTC2 cells, in the absence and presence of estrogen. 50 μg of <u>Xenopus</u> vitellogenin B1 gene-CAT [pB1(-596/8+)CAT8+] and 50 μg of either chicken ER expression construct [pMT2CER] or pMT2 alone were co-electroporated into a 0.5 ml volume of $1 \times 10^7$ XTC2 cells. After electroporation, the cells were divided into two equal portions, and these were plated in the presence ($+E_2$) or absence ($-E_2$) of estrogen ($E_2$). The cell extracts were then harvested 24 hr post-transfection. Lanes: 1, no extract added; 2-5, XTC2 cells transfected with pB1(-596/+8)CAT8+; 2 & 3, co-transfected with pMT2CER; 4 & 5, co-transfected with pMT2. The <u>Xenopus</u> vitellogenin B1 gene-CAT construct was kindly given by Prof. W. Wahli (Seiler-Tuyns et al., 1986), the chicken estrogen receptor was obtained from Prof. P. Chambon (Krust et al., 1986) and the vector pMT2 was from Genetics Institute.

6 also shows (lanes 1 and 2) that the ERE's in the promoter region are essential for hormonal response. Seiler-Tuyns et al. (personal communication) observed in a <u>Xenopus</u> kidney cell line which lacks ER that whereas the co-transfected human ER cDNA was able to promote the transcription of exogenous vitellogenin promoter-CAT constructs or of vitellogenin minigenes but failed to activate the endogenous chromosomal vitellogenin genes. Thus, primary cultures may offer a more suitable cellular milieu to study the sequence requirements for the regulation of hormone-mediated genes.

It is important that the receptor-gene interactions be also studied, in parallel with cell transfection, with transcription of cloned DNA sequences in cellular and nuclear extracts, DNA-protein complex formation by electrophoretic gel retardation assay and by DNase foot-printing. Such studies carried out with various genes regulated by different steroid hormones show in general that, besides the receptor, other tissue-specific transcription factors must also play a role in the expression of these genes.

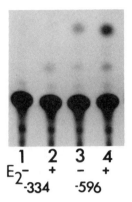

Figure 6. CAT assay of <u>Xenopus</u> primary hepatocytes transfected with
vitellogenin Bl-CAT constructs (-334) and (-596) with and without the
addition of estradiol ($E_2$). Lanes: 1, pBl (-334/+8)CAT8+, without $E_2$;
2, pBl(-334/+8)CAT8+, with $E_2$; 3, pBl(-596/+8)CAT8+, without $E_2$; 4,
pBl(-596/+8)CAT8+, with $E_2$. $1 \times 10^7$ cells were transfected with 100 μg
of DNA (lanes 1-4). Other conditions as in Figure 5.


## Tissue Specificity of Gene Expression

A characteristic feature of the action of steroid hormones, and other
hormones that act through mechanisms via nuclear receptors, is the high
degree of tissue specificity, i.e. different gene products are expressed
in different tissues in response to the same hormonal signal (Tata,
1984). Since no tissue-specific differences have yet been recorded for
the structure and function of steroid receptors, it is important to
consider putative tissue-specific factors that act in concert with
hormone receptors to give rise to differential gene expression. With
this goal in mind, we decided to explore another estrogen-inducible
gene(s) expressed selectively in a non-hepatic tissue of the <u>Xenopus</u>.
Since the oviduct is a major target for estrogen in all oviparous
vertebrates, we decided to initiate studies on <u>Xenopus</u> oviduct with the
long-term goal of obtaining tissue- and hormone-specific gene probes
that would enable us to compare the expression of vitellogenin and an
estrogen-inducible, oviduct-specific gene in cell transfection and in
<u>in vitro</u> DNA-protein interaction studies.


## Regulation of Gene Expression by Estrogen in the Xenopus Oviduct

Until recently, there was virtually no information on the genes that are
specifically expressed in the amphibian oviduct and their possible
control by estrogen. We therefore initiated work on proteins secreted
by the <u>Xenopus</u> oviduct, their possible role as egg proteins and the
expression of genes encoding these proteins (Maack et al., 1985; James
et al., 1985). More recently, we developed an estrogen responsive
<u>Xenopus</u> oviduct primary cell culture system which would enable us to
follow the kinetics of hormonal induction of specific mRNAs, even if the
nature of the proteins these mRNAs encode were unknown (Marsh and Tata,
1987a,b).

**L Oo Ov**

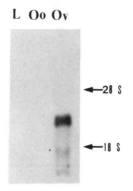

←—28 S

←—18 S

Figure 7. Autoradiogram of Northern blot hybridization of RNA from female Xenopus liver, oocytes and oviduct with [32]P-labelled FOSP-1 cDNA. Total RNA from the three tissues was electrophoresed on 1.5% agarose gels, transferred by blotting onto Gene Screen membrane and hybridized with plasmid pFOSP-1 labelled with [32]P. After washing, the filters were autoradiographed. L: liver; Oo: oocytes; Ov: oviduct. (From Lerivray et al., 1988).

## Cloning and Characterization of cDNA to Frog Oviduct Specific Protein 1 (FOSP-1)

Estrogen enhances the secretion of jelly proteins by the Xenopus oviduct, some of which are recognized by monoclonal antibodies raised against Xenopus egg jelly coats (Maack et al., 1985). Using a short cloned cDNA coding for an RNA induced by estrogen in Xenopus oviduct (James et al., 1985) as a screening probe, we have recently isolated a cDNA clone specifying an estrogen-inducible protein termed FOSP-1 (Lerivray et al., 1988).

That FOSP-1 cDNA indeed specified a gene exclusively expressed in oviduct was verified by Northern blot analysis of RNA extracted from different Xenopus tissues. Figure 7 shows that FOSP-1 cDNA does not hybridize to RNA in estradiol-treated female Xenopus liver and in oocytes but gives a strong signal with RNA from oviduct. Sequence analysis of 60% of FOSP-1 cDNA showed that only approximately half the length of FOSP-1 mRNA would include the coding sequence. A computer search of amino acid sequence homology in open reading frame revealed no homology with any protein sequence deposited in the data bank. Worth noting was the clustering of 10 basic amino acids (Arg and Lys) out of 12 at position 53-64, three asparagines at position 33-39 and a run of seven glutamines at the carboxyl end of the partial sequence. Southern blot analysis demonstrated that FOSP-1 gene does not belong to a multigene family but that it is most likely a single copy gene.

## Regulation of Expression of FOSP-1 Gene by Estrogen in Primary Cultures of Xenopus Oviduct Cells

For the same reasons mentioned above (see p. ) for studying the regu-lation of expression of vitellogenin genes in culture, rather than in

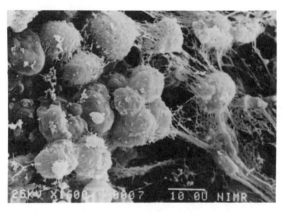

Figure 8. Scanning electron micrograph of <u>Xenopus</u> oviduct cells in
culture. Oviduct cells were plated at a density of 1.5 x $10^8$ cells/$cm^2$
and left for 3 days, the medium being changed daily, before they were
prepared for electron microscopy. Note the tendency of the cells to
clump together and the large amounts of jelly-like material secreted.
(From Marsh and Tata, 1987a.)

whole animals, it was decided to follow the regulation by estrogen of
FOSP-1 gene expression in primary cultures of <u>Xenopus</u> oviduct cells.
Despite serious problems caused by the structure of the amphibian
oviduct and jelly material stored in its lumen, we succeeded in establishing a system of oviduct primary cultures (Marsh and Tata, 1987a).
These cells, shown in Figure 8, are very active in secreting large
amounts of jelly proteins into the culture medium and retained their
response to estrogen. Addition of estrogen to the cultures enhanced the
synthesis and secretion of total protein by these cells. Estrogen
receptor in these cells had the same characteristic as that in <u>Xenopus</u>
liver and in all other estrogen target tissues (kd of $\sim$ 5 x $10^{-10}$ M).
The receptor content, after 4 days in culture, was lower than in avian
oviduct (Palmiter et al., 1978), but about the same as in female <u>Xenopus</u>
liver (Perlman et al., 1984). The establishment and characterization of
an oviduct culture system, comparable to primary cultures of
hepatocytes, is among the first successful attempts reported.

Having thus established the validity of our primary <u>Xenopus</u> oviduct cell
cultures, it became possible to study specifically the modulation by
estrogen of FOSP-1 gene in culture (Lerivray et al., 1988). The time-
course of accumulation of FOSP-1 mRNA in oviduct cell cultures following
the addition of estradiol-17ß is shown in Figure 9. Within 12 hr after
the addition of the hormone there was an 8-fold increase in the
concentration of FOSP-1 mRNA, reaching values 20 times higher than
control at 48 hr and plateauing at 25-fold at 96 hr. The concentration
of actin mRNA in the same RNA samples remained unchanged throughout this
period. Dose-response studies showed that the effect of estradiol
plateaued at 5 x $10^{-8}$ M, half-maximum values being reached at about 8 x
$10^{-10}$ M which demonstrates that in stimulating the expression of FOSP-1
gene the hormone is acting via the estrogen receptor. Progesterone
was only slightly active, raising the mRNA content to 2-fold at $10^{-7}$ M.
These results demonstrate that the FOSP-1 gene is expressed both in a
tissue- and hormone-specific manner.

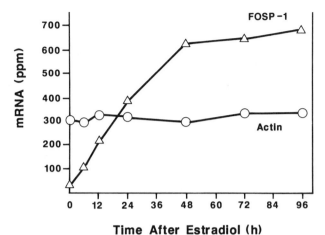

**Time After Estradiol (h)**

Figure 9. Time-course of accumulation of FOSP-1 mRNA in Xenopus oviduct cells in response to estrogen. After 3 days in culture, oviduct cells were exposed to $10^{-7}$M estradiol-17ß. At the different times indicated up to 96 hr after the addition of the hormone, RNA was extracted from batches of cells and used for determination of the concentration of FOSP-1 (△) and cytoskeletal actin (O) mRNAs. (From Lerivray et al., 1988.)

We are currently attempting to obtain a full-length FOSP-1 cDNA, both with a view to completing the sequence information and to derive a protein expressed in bacteria, which would be used to raise antibodies. Immunological analysis would then allow us to determine whether or not the protein is destined for the developing egg, in which case it would be most important to determine whether or not it is one of the constituents of egg coats (Maack et al., 1985). At the same time, we are attempting to derive clones of FOSP-1 from a Xenopus genomic library. These would enable us to compare the putative estrogen response elements with those of other Xenopus estrogen target genes, such as vitellogenin (Martinez et al., 1987; Klock et al., 1987), as well as with other steroid response elements in other organisms (Green et al., 1987; Becker et al., 1987). The cloning and identification of the promoter and other regions of the genes will enable us to analyse the fundamental question of tissue specificity both by transfection of DNA constructs into cultured cells and analysis in cell-free systems of transcription and protein-DNA interactions.

### Developmental Acquisition of Competence to Respond to Estrogen

How early in development is this pattern of differential activation within a gene family seen in adult cells established? We found that Xenopus tadpole hepatocytes acquired competence to synthesize vitellogenin mRNA in response to the hormone by Nieuwkoop-Faber stage 61, i.e. estrogen receptor was present in hepatocytes at least by late metamorphosis (Ng et al., 1984). Measurement of individual vitellogenin gene transcripts in metamor-phosing tadpole liver revealed the same relative pattern of expression as in adult hepatocytes, i.e. gene B1 > A1 > A2 ≃ B2, at the earliest stages of activation of these dormant

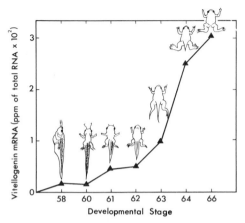

Figure 10. Acquisition of competence to express vitellogenin genes in response to estrogen during Xenopus development, as determined by the accumulation of total vitellogenin mRNA. Xenopus laevis tadpoles of various developmental stages were maintained in tap water containing $10^{-6}$M 17ß-estradiol for a period of 3 days before the accumulation of total vitellogenin mRNA specified by all four genes in tadpole liver was measured by hybridization to an equimolar mixture of the plasmids containing the four vitellogenin cDNAs. The developmental stages used are diagrammatically represented on the graph. (From Ng et al., 1984.)

genes. Thus, the unequal pattern of expression is maintained throughout life, although the absolute rate of transcription of each gene increases rapidly between late metamorphic and froglet stages.

One of the mechanisms determining the ability of tadpole liver to start expressing the dormant vitellogenin genes in response to the hormone would be the acquisition of functional estrogen receptor. Indeed, measurement of receptor content in metamorphosing tadpole liver confirmed this assumption (May and Knowland, 1981). It is, however, important to note that the onset and build-up of receptor could also be accompanied by alterations in the chromatin configuration of vitellogenin genes, as observed for vitellogenin genes in developing chick embryonic liver (Burch and Weintraub, 1983).

Since estrogen receptor belongs to a superfamily of genes related to the proto-oncogene c-erb-A, which also encodes receptors for other steroid hormones, thyroid hormone and retinoic acid (Shepel and Gorski, 1988; Evans, 1988), we thought it would be of some interest to measure the amount of c-erb-A transcripts during Xenopus development. In results obtained recently in our laboratory (Baker and Tata, unpublished), we found that there was a continuous increase in the accumulation of c-erb-A mRNA in parallel with the acquisition during late metamorphosis of vitellogenin response to estrogen (Figure 10). The pattern for c-erb-A transcripts, shown in Figure 11, represents a second burst of accumulation of the mRNA, since it was preceded at early tadpole development (stages 40-42) by a transient burst. Interestingly, this first accumulation coincided with the early acquisition of competence to respond to thyroid hormones which regulate metamorphosis at later

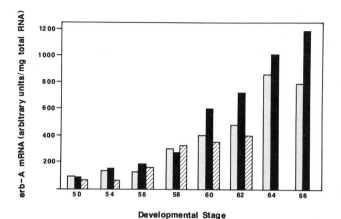

**Developmental Stage**

Figure 11. Accumulation of c-<u>erb</u>-A transcripts in late <u>Xenopus</u> tadpole
stages and during spontaneous metamorphosis. Tadpoles were dissected
into the head (▨), middle (■), and tail (▨) regions and total RNA
extracted separately from these. Due to extensive tissue regression and
cell death at these stages of metamorphosis, RNA was not extracted from
tail remnants at stages 64 and 66. The relative amounts of c-<u>erb</u>-A
transcripts in these RNA samples was determined by quantitative slot-
blot hybridization.

developmental stages. Thus, it may well turn out that the bursts of
accumulation of transcripts of ancestral oncogenes from which hormone
and growth factor receptors have evolved reflect an overall strategy of
intercellular communication networks being established during
development.

## General Conclusions

Hormonally coordinated development of the egg is an exquisite example of
division of labor for producing specialized gene products, essential for
ensuring reproduction of the organism. Whereas the direct involvement
of the follicle cell in providing egg components remains unknown, the
regulation by estrogen of egg protein gene expression in the liver and
oviduct of oviparous vertebrates has been well defined and intensively
studied. Recently, with the exploitation of techniques of gene cloning
and transfection, <u>Xenopus</u> vitellogenin genes constitute an increasingly
popular system for determining the fine details of gene activity as well
as the action of steroid hormones.

The use of primary cultures of male and female <u>Xenopus</u> hepatocytes in
our laboratory has allowed a more precise analysis of the early events
accompanying the <u>de novo</u> activation of the dormant vitellogenin genes.
For example, the demonstration of unequal rates of transcription of the
four vitellogenin genes points to the importance of the interaction
**between estrogen receptor and ERE's and other transcription regulatory**
sequences. An interesting feature of estrogen receptor in male <u>Xenopus</u>
hepatocytes is its up-regulation by estrogen itself from very low
levels. The small amount of estrogen receptor in male hepatocytes has

also made it possible to define an almost stoichiometric relationship between receptor number and the absolute rate of transcription of the induced gene in vivo.

An important question to resolve in further understanding the regulation of gene expression by steroid hormones is the molecular basis of specificity of different genes regulated in different target tissues. It is for this reason that we initiated work on characterizing protein and mRNA products of genes expressed exclusively in Xenopus oviduct and inducible with estrogen. The results of our studies on the expression of the oviduct-specific, estrogen-inducible mRNA of FOSP-1 bring us closer towards the goal of unravelling tissue specificity. The kinetics of induction of FOSP-1 mRNA in oviduct cell cultures by estrogen and the dose-response curves resemble those seen earlier for transcription of vitellogenin genes in male and female hepatocyte cultures. This most likely indicates the key role of estrogen receptor, identical in both tissues, so that the qualitative difference as to which gene is regulated in one cell type or another must originate from some tissue-specific factors. The elucidation of the nature of such factors, their interaction with specific DNA sequences and with other transcription regulatory elements will turn out to be of utmost importance in furthering our understanding of tissue specificity of differential gene expression, a key feature of maintenance of differentiated states of eukaryotic cells.

Another question of wider significance is how the hormonal and developmental regulation of expression of egg protein genes fits into the overall scheme of intercellular communication via external signals and their intracellular transduction. In this context, it is of some significance that receptors for many extracellular signals such as hormones, growth factors, morphogens, etc. are products of multigene families related to cellular oncogenes. Thus, steroid and non-steroid hormones and morphogens acting via nuclear receptors and peptide hormones, growth factors and other signals acting through cell membrane receptors fall into the erb-A and erb-B oncogene superfamilies, respectively. Hormonal and developmental coordination of egg protein genes should therefore be viewed as one of the physiological functions dependent on a tightly integrated network of intercellular communication.

## Acknowledgements

We wish to thank Mrs. Betty Baker and Mrs. Jacky Smith for their participation in some of the work described in this article. We also thank Mrs. Ena Heather for preparation of the manuscript.

## References

Baker, H.J. and Shapiro, D.J. J. Biol. Chem. 252, 8428-8434, 1977.
Becker, P.B., Ruppert, S. and Schutz, G. Cell 51, 435-443, 1987.
Brock, M.L. and Shapiro, D.J. Cell 34, 207-214. 1983.
Browder, L.W. (ed.). Developmental Biology, Oogenesis. Vol. 1, Plenum Press, New York. 1985.
Burch, J.B.E. and Weintraub, H. Cell 33, 65-76, 1983.
Chambon, P., Dierich, A., Gaub, M.-P., Jakowlev, S., Jongstra, J., Krust, A., LePennec, J.-P., Oudet, P. and Reudelhuber, T. Recent Prog. Horm. Res. 40, 1-39, 1984.

Chu, G., Hayakaw, H. and Berg, P. Nucl. Acids Res. 15, 1311-1326, 1987.
Dean, D.C., Gope, R., Knoll, B.J., Riser, M.E. and O'Malley, B.W.
  J. Biol. Chem. 259, 9967-9970, 1984.
Eriksson, J. and Gustafsson, J.-A. Steroid Hormone Receptors: Structure
  and Function. Elsevier, Amsterdam. 1983.
Evans, R.M. Science 240, 889-895, 1988.
Farmer, S.R., Henshaw, E.C., Berridge, M.V. and Tata, J.R.   Nature 273,
  401-403, 1978.
Germond, J.-E., Walker, P., Heggeler, T.B., Brown-Luedi, M., de Bony,
  E. and Wahli, W. Nucl. Acids Res. 12, 8595-8609, 1984.
Green, S. and Chambon, P. Nature 325, 75-78, 1987.
Green, S., Kumar, V., Krust, A. and Chambon, P.  In: Recent Advances in
  Steroid Hormone Action.  (V.K. Moudgil, ed.) pp. 161-183.  Walter de
  Gruyter, Berlin, 1987.
Hayward, M.A., Mitchell, T.A. and Shapiro, D.J. J. Biol. Chem. 255,
  11308-11312, 1980.
James, T.C., Maack, C.A., Bond, U.M., Champion, J. and Tata, J.R.
  Comp. Biochem. Physiol. 80B, 89-97, 1985.
Jost, J.-P., Seldran, M. and Geiser, M.  Proc. Natl. Acad. Sci. USA 81,
  429-433, 1984.
Klein-Hitpass, L., Schorpp, M., Wagner, U. and Ryffel, G.U.
  Cell 46, 1053-1106, 1986.
Klock, G., Strahle, U. and Schutz, G. Nature 329, 734-736, 1987.
Krust, A., Green, S., Argos, P., Kumar, V., Walter, P., Bornet, J.-M.
  and Chambon, P.  EMBO J. 5, 891-897, 1986.
Lerivray, H., Smith, J. and Tata, J.R. Mol. Cell. Endocrinol. 1988
  in press.
Maack, C.A., James, T.C., Champion J., Hunter, I.R. and Tata, J.R.
  Comp. Biochem. Physiol. 80B., 77-87, 1985.
Marsh, J. and Tata, J.R.   Exp. Cell Res. 173, 117-128, 1987a.
Marsh, J. and Tata, J.R.   Mol. Cell. Endocrinol. 53, 141-148, 1987b.
Martinez, E., Givel, F. and Wahli, W. EMBO J. 6, 3719-3727, 1987.
May, F.E.B. and Knowland, J. Nature 292, 853-855, 1981.
Metz, C.B. and Monroy, A. (eds.).  Biology of Fertilization.  Vol. 1,
  Plenum Press, New York.  1985.
Migliaccio, A., Rotondi, A. and Auricchio, F. EMBO J. 5, 2867-2872,
  1986.
Ng, W.C., Wolffe, A.P. and Tata, J.R. Dev. Biol. 102, 238-247, 1984.
Palmiter, R.D., Mulvihill, E.R., McKnight, G.S. and Senear, A.W.
  Cold Spring Harbor Symp. Quant. Biol. 42, 639-647, 1978.
Payvar, F., DeFranco,D., Firestone,G.L., Edgar, B., Wrange, O., Okret,
  S., Gustafsson, J.-A. and Yamamoto, K.R.   Cell 35, 381-392, 1983.
Perlman, A.J., Wolffe, A.P., Champion, J. and Tata, J.R. Mol. Cell.
  Endocrinol. 38, 151-161, 1984.
Renkawitz, R., Schutz, G., Van der Ahe, D. and Beato, M.   Cell 37,
  503-510, 1984.
Schubiger, J.-L. and Wahli, W. Nucl. Acids Res. 14, 8723-8734, 1986.
Searle, P.F. and Tata, J.R. Cell 23, 741-746, 1981.
Seiler-Tuyns, A., Walker, P., Martinez, E., Merillat, A.-M., Givel, F.
  and Wahli, W. Nucl. Acids. Res. 14, 8755-8770, 1986.
Shepel, L.A. and Gorski, J. BioFactors 1, 71-83, 1988.
Tata, J.R.  In: Biological Regulation and Development, Vol. 3B,
  (R.F. Goldberger and K.R. Yamamoto, eds.) pp. 1-58, Plenum Press,
  New York.  1984.
Tata, J.R.  In: The Endocrine System and the Environment. (B.K. Follett,
  S. Ishii and A. Chandola, ed.) pp. 85-91, Japan Sci. Soc. Press,
  Tokyo/Springer-Verlag, Berlin.  1985.
Tata, J.R.  BioEssays 4, 197-201. 1986.
Tata, J.R. In: Recent Advances in Steroid Hormone Action, (V.K. Moudgil,
  ed.) pp. 259-287, Walter de Gruyter & Co., Berlin/New York, 1987.

Tata, J.R. In: Developmental Biology, A Comprehensive Synthesis, Vol. 5, (L.W. Browder, ed.) Plenum Publishing, New York. pp.241–265. 1988.

Tata, J.R. and Smith, D.F. Recent Prog. Horm. Res. 35, 47-95, 1979.

Tata, J.R., A.P. Wolffe, A.J. Perlman and W.C. Ng. In: Nuclear Envelope Envelope Structure and RNA Maturation, UCLA Symposia on Molecular and Cellular Biology, New Series, Vol. 26, (E.A. Smuckler and G.A. Clawson, eds.) pp. 357-378, Alan R. Liss, New York. 1985.

Tur-Kaspa, R., Teicher, L., Levine, B.J., Skoultchi, A.I. and Shafritz, D.A. Mol. Cell. Biol. 6, 716-718, 1986.

Wahli, W., Dawid, I.B., Wyler, T., Jaggi, R.B., Weber, R. and Ryffel, G.U. Cell 16, 535-549, 1979.

Wahli, W., Germond, J.-E., Heggeler, B.T. and May, F.E.B. Proc. Natl. Acad. Sci. USA 79, 6832-6836, 1982.

Walker, P., Germond, J.-E., Brown-Luedi, M., Givel, F. and Wahli, W. Nucl. Acids Res. 12, 8611-8626, 1984.

Wallace, R.A. In: Developmental Biology, Oogenesis. Vol. 1, (L.W. Browder, ed.) pp. 127-177, Plenum Press, New York. 1985.

Weiler, I.J., Lew, D. and Shapiro, D.J. Mol. Endocrinol. 1, 355-362, 1987.

Westley, B. and Knowland, J. Cell 15, 367-374, 1978.

Westley, B. and Knowland, J. Biochem. Biophys. Res. Commun. 88, 1167-1172, 1979.

Wolffe, A.P. Ph.D. Thesis. Council for National Academic Awards, London. 1984.

Wolffe, A.P. and Tata, J.R. Eur. J. Biochem. 130, 365-372, 1983.

Wolffe, A.P. and Tata, J.R. FEBS Lett. 176, 8-15, 1984.

DISCUSSION OF THE PAPER PRESENTED BY J.R. TATA

ROY: Just to have a bit more about the transfection in the primary culture, I assume you are using the BioRad electroporation apparatus. Would you be kind enough to provide more details about the tid-bits of the procedure (if these are not published yet)?

TATA: Primary cell cultures, suspended in L15 medium at a density of $1 \times 10^6 - 1 \times 10^7$ cells/0.5 ml, were incubated for 10 min. with 1-100 µl/ml of the appropriate supercoiled DNA. The cell-DNA suspension was placed in a Bio-Rad type of electroporation chamber fitted with platinum or aluminum foil electrodes. The electric pulse, usually at 600-800 v, was discharged across the cell suspension from a 800 µF capacitor. After electroporation, the cells were diluted with culture medium to different cell densities and incubated generally for 48 h. before carrying out CAT assays or transcript analysis.

STANCEL: In your developmental studies, what marker were you using to assess sensitivity to thyroid hormone? Also, has anyone ever checked in Xenopus to see if $T_3$ stimulates production of EGF and/or EGF receptor, since we know that $T_3$ elevates EGF receptor in mammalian liver?

TATA: Various indicators of response to $T_3$ were used. The most sensitive response was the uptake of $^{32}PD_4$--- ions, but also overall RNA and protein synthetic activity were monitored. I am not aware of any studies on the effect of $T_3$ on EGF or EGF receptor in amphibia.

DISCUSSANTS: A.K. ROY, J.R. TATA, G. STANCEL

Inhibition of the Intracellular Transformation of Rat Ventral Prostate
Androgen Receptor by 3'-Deoxyadenosine

Richard A. Hiipakka and Shutsung Liao

Introduction

Numerous studies have documented the effects of steroids on the transcription of
specific genes. These studies and the elucidation of the structure of steroid receptors
and steroid-responsive genes have supported the hypothesis that steroid-receptor
complexes affect gene transcription by acting as transcriptional regulators that interact
with gene regulatory elements (Yamamoto). Although steroids have dramatic effects
on the transcription of certain genes, there is evidence for post-transcriptional effects
of steroids on the expression of certain gene products (Shapiro et al.,1985,1987).
For example, in chick oviduct the level of a mRNA for a heat shock protein increases
20 to 50-fold with estrogen or progestin treatment, while the rate of transcription of
the gene is increased only 2 to 4-fold (Baez et al.). In the rat ventral prostate,
androgens enhance the transcription of the genes for the subunits of a steroid-binding
secretory protein about 2 to 3-fold; however, the level of the mRNAs for this protein
increase about 30-fold (Page et al.). Glucocorticoids and thyroid hormones increase
the transcription of the growth hormone gene in rat pituitary cells about 5-fold, but
the level of the mRNA increases 50-fold (Diamond et al.). Therefore, steroids affect
not only the transcription of genes but also alter the stability of the induced mRNA.
The mechanism of this increased stability is not known.

We have been investigating the possibility that steroid receptors, as part of the
mechanism by which steroids control gene expression, bind RNA transcripts in the
nucleus and affect various aspects of the processing, transport, stability and
utilization of RNA (Liao et al., 1969,1972,1973). We have shown that androgen-
and estrogen-receptor complexes bind RNA or ribonucleoprotein in the rat ventral
prostate (Liao et al., l973) or uterus (Liang et al.). We have also shown that, in a
cell-free system, certain polyribonucleotides can bind steroid-receptor complexes
and promote the release of the complexes from DNA and that this effect is dependent
on the nucleotide sequence of the RNA (Liao et al., 1980). Similar results have been
presented by other investigators for various steroid-receptor systems (Feldman et al.;
Lin et al.; Franceschi; Anderson et al.; Ali et al.; Mulder et al.). Treatment of various

steroid-receptor complexes with RNase increases binding of receptors to DNA and changes their sedimentation properties (Chong et al.; Tymoczko et al.; Rossini; Rowley et al.). These studies demonstrate that steroid receptors have the potential to interact with RNA. If such an interaction occurs in a cell, steroid receptors may be directly responsible for some of the post-transcriptional effects of steroids on specific RNAs.

The interaction of steroid receptors with RNA may be part of a recycling mechanism which controls the distribution of chromatin-bound and cytosolic forms of steroid receptors (Rossini et al.). Interference with the recyling process may alter the distribution of steroid receptors in a cell. Incubation of prostatic tissue with actinomycin D, an inhibitor of RNA synthesis, increases chromatin-bound and decreases cytosolic levels of androgen receptors (Rossini et al.). Actinomycin D also increases or stabilizes binding of estrogen receptors to nuclear chromatin of rat uterus (Schoenberg et al.) and MCF-7 cells (Horwitz et al.). Inhibition of RNA synthesis by actinomycin D may prevent release of androgen receptors, increasing or stabilizing chromatin-bound receptor levels, if synthesis of RNA and recycling of receptors are coupled. As a further test of the role of RNA in receptor recycling we have studied (Hiipakka et al.) the effects of 3'-deoxyadenosine (3'-dA), an inhibitor of the synthesis (Beach et al.), polyadenylation (Darnell et al.), and nucleocytoplasmic transport (Agutter et al.; Kletzein) of RNA. If androgen receptors are involved in these processes as part of a recycling mechanism, 3'-dA may interfere with recycling of androgen receptors and interaction of androgen receptors with nuclear chromatin or DNA.

Experimental Procedure

To study the effects of 3'-dA on prostatic androgen receptors we used a short term culture of ventral prostate tissue from rats castrated 18 to 24 h previously ( Rossini et al.) The ventral prostates from 6-10 rats were pooled, minced, and mixed with Dulbecco's modified Eagle's medium (DMEM). Tissue (0.3-0.5 g) was incubated in a final volume of 5 ml of DMEM (equilibrated with 95% $O_2$ ,5% $CO_2$) containing 20 mM HEPES, pH 7.0, the indicated amounts of nucleoside, and 10 μM erythro-9-(2-hydroxy-3-nonyl)adenine (EHNA), an inhibitor of adenosine deaminase, which potentiates the effect of 3'-dA (Plunkett et al.). Incubations were performed at 37 °C for 30 min with constant shaking in a water bath. [$^3$H]Dimethylnortestosterone (DMNT), 75 Ci/mmole, in DMEM was then added to a final concentration of 10 nM and the incubations were continued for an additional 30 min and then placed on ice. Tissue was collected by centrifugation, washed once with 0.32 M sucrose containing 1 mM $MgCl_2$, and 20 mM Tris-HCl, pH 7.5, homogenized, and separated into cytosolic and nuclear fractions by centrifugation at 3500 x g for 10 min. Androgen receptors in the cytosolic fraction were assayed by a hydroxylapatite assay (Liao et al., 1984). The ability of cytosolic androgen receptors to bind to DNA-cellulose was determined (Liao et al., 1980) and labeled androgen receptors in the nuclear fraction were determined after extraction with buffer containing 0.5 M KCl (Rossini et al.). Results are expressed as the number of [$^3$H]DMNT-labeled androgen-receptor complexes per prostate cell based on a determination of the DNA content (Burton) of the nuclear fraction. Each diploid cell was assumed to contain 6 pg of DNA.

## Effect of 3'-dA on Prostatic Androgen Receptors

We first determined if 3'-dA had an effect on rat ventral prostate androgen receptors by incubating minced prostatic tissue with 1 mM 3'-dA for 30 min before and after a 30 min incubation with labeled steroid (table 1). We then analyzed the level of labeled androgen receptors in cytosolic and chromatin fractions. If labeled steroid was added prior to the addition of 3'-dA, no effect of 3'-dA was observed on the distribution of receptor. However, if 3'-dA was added 30 min before the addition of steroid, a reduction in the level of labeled receptor in the chromatin fraction and an increase in the level of receptor in the cytosolic fraction was found. As we have observed previously (Rossini et al.), incubations in which labeled steroid was added after an initial incubation without steroid resulted in tissue with a lower total receptor content compared to tissue from incubations where steroid was added at the beginning of the incubation. This difference may reflect a lability of unlabeled receptor in this system. The effect of the length of incubation time with 3'-dA, prior to addition of steroid, on chromatin-bound and cytosolic androgen receptors is given in figure 1.

The level of cytosolic androgen receptors increased and reached a plateau after 5 min of incubation with 1 mM 3'-dA. In contrast, the levels of chromatin-bound

Table 1. Effect of the order of incubation with 3'-dA and [$^3$H]DMNT on rat ventral prostate androgen receptor*.

| Addition | | Androgen receptor complexes (molecules/cell) | | |
|---|---|---|---|---|
| 1st Incubation | 2nd Incubation | Cytosol | Chromatin-bound | Total |
| None | [$^3$H]DMNT | 8515 | 5186 | 13701 |
| 3'-dA | [$^3$H]DMNT | 11608 | 1294 | 12902 |
| [$^3$H]DMNT | None | 10880 | 9877 | 20757 |
| [$^3$H]DMNT | 3'-dA | 9445 | 10465 | 19910 |

*Minced prostatic tissue was incubated first for 30 min with either [$^3$H]DMNT (10 nM) or 3'-dA (1 mM), then [$^3$H]DMNT or 3'-dA was added as indicated and the incubations continued for an additional 30 min. Labeled androgen receptors in the cytosolic and chromatin fractions were then determined. (Hiipakka and Liao, 1988)

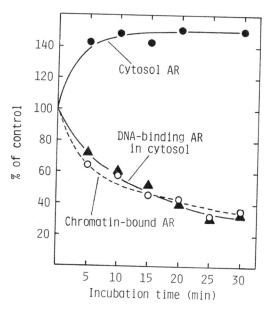

Fig. 1 Effect of incubation time with 3'-dA on the rat ventral prostate androgen receptor. Minced prostatic tissue was incubated with 1 mM 3'-dA for the time indicated, and then [³H]DMNT was added and the incubation continued for an additional 30 min. The amount of labeled androgen receptor in the cytosolic ( ● ) and chromatin ( o ) fractions and the DNA-cellulose binding capacity ( ▲ ) of the androgen receptor (AR) in the cytosolic fraction was determined. Results are expressed as a percent of the value from a control which was incubated without 3'-dA for the same length of time before addition of [³H]DMNT. (Hiipakka and Liao, 1988)

receptor slowly decreased reaching a minimum after 25-30 min of incubation with 3'-dA. Since the simultaneous increase in cytosolic receptor levels and decrease in nuclear receptor levels indicated a loss of chromatin-binding activity, we also analyzed the effect of 3'-dA on the DNA-cellulose binding activity of labeled cytosolic androgen receptors (figure 1). Incubation of tissue with 3'-dA also decreased the amount of cytosolic recptor that could bind to DNA-cellulose and this decrease paralleled the loss of chromatin-bound androgen receptor.

We next examined the effect of incubation with various concentrations of 3'-dA prior to addition of labeled steroid (figure 2). Incubation with 3'-dA led to a dose-dependent reduction in the level of chromatin-bound androgen receptor (50% decrease with 0.3 mM 3'-dA), an increase in the level of androgen receptor in the

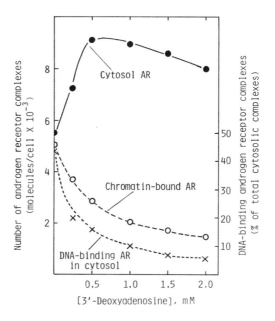

Fig. 2 Effect of concentration of 3'-dA on the rat ventral prostate androgen receptor. Minced prostatic tissue was incubated first with various concentrations of 3'-dA for 30 min, and then [³H]DMNT was added and the incubation continued for an additional 30 min. Cytosolic and nuclear fractions were then prepared from incubated tissue and the amount of [³H]DMNT-labeled androgen receptor (AR) in the cytosolic ( ● ) and chromatin ( o ) fractions and the amount of labeled androgen receptor binding to DNA-cellulose (in % of total cytosolic androgen receptors) (x) was determined. (Hiipakka and Liao, 1988)

cytosolic fraction, and a decrease the amount of cytosolic androgen receptor that was able to bind to DNA-cellulose. Since the decrease in chromatin-bound receptor and DNA-cellulose-binding activity may be due to increased binding of RNA to receptors, we treated labeled cytosolic androgen receptors from 3'-dA-treated tissue with pancreatic RNase A. RNase A treatment increased labeled androgen receptor binding to DNA-cellulose by only 10-20%. Increasing the amount of RNase or the time or temperature of incubation did not increase the binding activity further. Therefore, either RNA was absent from the receptor, not accessible to RNase A, or removal of RNA was not sufficient to restore DNA-binding activity. Other methods that can transform steroid receptors to a DNA-binding state, such as ammonium sulfate precipitation or heating did not increase binding of labeled cytosolic androgen receptors to DNA-cellulose. Untransformed estrogen receptors will bind to DNA after exposure to high concentrations of urea (Hutchens et al.). Therefore, we treated cytosol from tissue incubated with 1 mM 3'dA with 5 M urea which increased DNA-cellulose binding of [³H]DMNT-labeled androgen receptor 3.5-fold. This represents a 70% recovery of the DNA-binding activity of the cytosolic androgen receptor complexes which was lost during the incubation of prostatic tissue with 3'dA.

We tested the nucleoside specificity of the 3'-dA effect using various other nucleosides, such as adenosine, 2'-deoxyadenosine, cytidine, 2'-deoxycytidine, guanosine, 2'-deoxyguanosine, 2'-deoxyinosine, 5'-deoxythymidine and uridine; only 3'-dA affected the subcellular distribution of androgen receptors. Since 3'-dA may be competing with certain nucleosides, we also tested the ability of various nucleosides to inhibit the effect of 3'-dA when added to tissue incubations along with 3'-dA. Cytidine, guanosine or uridine had little or no effect when added at concentrations up to 10 mM and more than 10 times the concentration of 3'-dA. Adenosine, however, at a concentration of 10 mM completely inhibited the effect of 1 mM 3'-dA. Adenosine was a competitive inhibitor of the 3'-dA effect (figure 3). Therefore, a process requiring adenosine or one of its metabolic products may be affected by 3'-dA. Adenosine may also compete with 3'-dA for transport into prostatic tissue and may inhibit the effect of 3'-dA in this manner. Since an incuba-

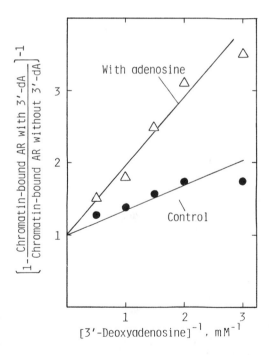

Fig. 3 Kinetic analysis of adenosine inhibition of the effect of 3'-dA on chromatin-bound androgen receptor. Minced prostatic tissue was incubated with the indicated concentration of 3'-dA with ( $\Delta$ ) or without ( $\bullet$ ) 1 mM adenosine for 30 min, and then [$^3$H]DMNT was added and the incubation continued for an additional 30 min. The amount of androgen receptor in the chromatin fraction was then determined. (Hiipakka and Liao, 1988)

tion period with 3'-dA, prior to labeling with [³H]DMNT, was required for the effect of 3'-dA on prostatic androgen receptors, the effect may also require metabolic conversion of 3'-dA to the nucleoside triphosphate. Inhibition of synthesis (Beach et al.), polyadenylation (Darnell et al.) , and nucleocytoplasmic transport (Agutter et al.; Kletzein) of RNA by 3'-dA requires conversion of this nucleoside to the nucleoside triphosphate which competes with ATP.

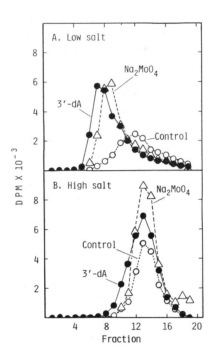

Fig. 4  Effect of 3'-dA on the sedimentation behavior of rat ventral prostate androgen receptor.  Minced prostatic tissue was incubated with ( ● ) or without  ( o ) 1 mM 3'-dA for 30 min, and then  [³H]DMNT was added and the incubation continued for an additional 30 min.  A cytosolic fraction was prepared from tissue in these incubations as well as a cytosolic fraction from minced tissue incubated without 3'-dA at 0° C for 60 min and homogenized in buffer containing 10 mM Na₂MoO₄ ( Δ ). This cytosolic fraction was labeled with 10 nM [³H]DMNT for 1 h at 0° C before analysis.  All cytosolic fractions were treated with dextran-coated charcoal before analyzing 0.2 ml samples on 3.8 ml 5-20% sucrose gradients containing 20 mM Tris-HCl,  pH 7.5, 1 mM EDTA, 1 mM dithiothreitol, 10% glycerol and without (A) or with (B) 0.4 M KCl.  Centrifugation was at 257,000 x g for 16 h at 4°C. Gradients were fractionated from the bottom and 0.2 ml per fraction collected.  The marker proteins, ovalbumin (3.5S), rabbit IgG (7S) and bovine catalase (11S), were run on parallel gradients for determination of sedimentation coefficients (Martin et al.) of the various forms of rat ventral prostate androgen receptors.  (Hiipakka and Liao, 1988)

The effect of 3'-dA appeared to be an inhibition of the intracellular transformation of androgen receptor to a chromatin-binding form. Besides lacking high affinity for chromatin or DNA, untransformed steroid receptors have larger sedimentation coefficients and increased affinity for anion-exchange resins, such as DEAE-cellulose compared to transformed receptors (Schmidt et al.). Therefore, we tested the effect of incubation of tissue with 3'-dA on prostatic androgen receptor binding to DEAE-cellulose and sedimentation in sucrose gradients.

In gradients containing 0.4 M KCl, labeled cytosolic androgen receptors had sedimentation coefficients of 4 S, regardless of whether tissue had been incubated with or without 3'-dA (figure 4). In low-salt gradients, the labeled cytosolic receptor from the control tissue had a broad peak with a sedimentation coefficient of 5.5 S. Labeled receptor from tissue treated with 3'-dA had a sharp peak with a sedimentation coefficient of 8-9 S. Treatment of cytosol from 3'-dA-treated tissue

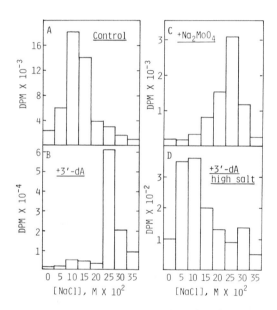

Fig. 5  Effect of 3'-dA on the binding properties of androgen receptor to DEAE-cellulose. Cytosolic fractions were prepared from minced prostatic tissue incubated without (A) or with (B) 1 mM 3'-dA, and from tissue not incubated with 3'-dA but homogenized in buffer containing 10 mM $Na_2MoO_4$ (C) and then labeled with [$^3$H] DMNT as described in figure 4. Charcoal-treated labeled cytosolic fractions (3 ml) were placed on columns of DEAE-cellulose (1.5 x 3 cm). A cytosolic fraction from tissue incubated with 3'-dA was also fractionated on a high-salt sucrose gradient, desalted on a Sephadex PD-10 column and placed on a DEAE-cellulose column (D). The columns were then eluted with step gradients (10 ml each) containing the indicated concentration of NaCl. The amount of labeled androgen receptor in each fraction was determined by the hydroxylapatite assay. (Hiipakka and Liao, 1988)

with pancreatic RNase A or calf intestinal alkaline phosphatase did not change the sedimentation behavior of labeled androgen receptor in low-salt gradients.

The sedimentation properties of [$^3$H]DMNT-labeled androgen receptor from 3'-dA-treated tissue are typical of untransformed androgen receptors prepared from tissue homogenized in the presence of 10 mM Na$_2$MoO$_4$ and labeled with [$^3$H]DMNT after removal of nuclei (Traish et al.). For comparison, therefore, we prepared labeled androgen receptor in the presence of Na$_2$MoO$_4$. Molybdate-stabilized untransformed receptor had a sedimentation coefficient of 8-9 S in low-salt gradients with a shoulder in the region of 5.5 S (figure 4). Since the gradient did not contain molybdate, this shoulder may represent transformed receptor generated during centrifugation. In high-salt gradients, receptor prepared in the presence of molybdate had a peak sedimentation coefficient of 4 S, the same as receptor from control or 3'-dA-treated tissue. [$^3$H]DMNT-labeled androgen receptor extracted from chromatin had a sedimentation coefficient of 3.5 S in low and high-salt gradients and this value was unaffected by treatment of tissue with 3'-dA (data not shown).

The elution profiles from DEAE-cellulose of [$^3$H]DMNT-labeled cytosolic androgen receptor from control and 3'-dA-treated tissue are shown in figure 5. Labeled cytosolic receptor from the control tissue eluted over a broad range of salt concentrations, but the majority of the receptor eluted before 0.25 M NaCl. In contrast, labeled receptor from 3'-dA-treated tissue eluted at higher salt concentrations, between 0.25 and 0.35 M NaCl. The elution of labeled receptor prepared in the presence of 10 mM Na$_2$MoO$_4$ was similar to that of labeled receptor from 3'-dA-treated tissue, although elution occurred over a broader range of salt concentration.

The effects of 3'-dA on rat ventral prostate androgen receptor are consistent with an inhibitory effect of this nucleoside on the intracellular transformation of androgen receptors. Untransformed receptors have lower affinity for DNA, larger sedimentation coefficients, usually in the region of 8-9 S, and increased affinity for anion-exchange resins, such as DEAE-cellulose compared to transformed receptors (Schmidt et al.). All of these characteristics are present in the [$^3$H]DMNT-labeled androgen receptors from prostatic tissue treated with 3'-dA.

The mechanism of receptor transformation and the composition of the untransformed receptor have not been clearly established. Various procedures, such as heating, dilution, ammonium sulfate precipitation, exposure to high salt concentrations and gel filtration can cause transformation of steroid receptors (Schmidt et al.). These procedures may be removing components from the untransformed receptor or may be causing structural changes in the receptor. The evidence for structural changes during receptor transformation has been equivocal (Schmidt et al.; Mendel et al., 1987; Smith et al.). Transformation of glucocorticoid receptors by heating or ammonium sulfate precipitation is accompanied by the loss of a Mr: 90,000 heat shock protein (HSP 90) from the untransformed receptor (Sanchez et al.). The role of HSP 90 in receptor function is not yet established, but HSP 90 is found associated with different steroid receptors in a variety of tissues (Riehl et al.). Molybdate, which stabilizes receptors in the untransformed state, also stabilizes the interaction of HSP 90 with steroid receptors (Sanchez et al.).

The lack of an effect of 3'-dA on the sedimentation characteristics of cytosolic androgen receptor in high salt gradients may be due to separation during centrifugation of some component(s) from androgen receptor prepared from 3'-dA-treated tissue. This component(s) may be responsible for the changes in receptor behavior described above. We tested this possibility by isolating [³H]DMNT-labeled receptor from 3'-dA-treated tissue after centrifugation on a high-salt gradient. Peak fractions of [³H]DMNT were pooled and desalted by gel filtration and then analyzed for sedimentation behavior on low-salt sucrose gradients, DNA-cellulose binding, and elution behavior on DEAE-cellulose.

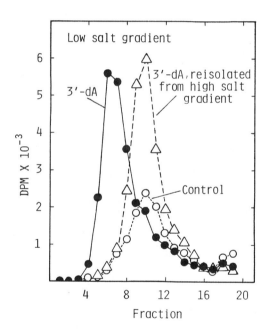

Fig. 6 Sedimentation analysis of androgen receptor from 3'-dA treated tissue before and after isolation on high-salt gradients. [³H]DMNT-labeled cytosol from tissue incubated with 3'dA was prepared and centrifuged on high-salt (0.4 M KCl) sucrose gradients. Peak fractions of radioactivity were pooled and desalted on a Sephadex PD-10 column. Peak fractions from the PD-10 column were pooled and an aliquot analyzed by low-salt sucrose gradient centrifugation ( Δ ) as described in figure 4. For comparison labeled cytosolic fractions prepared from tissue incubated with ( ● ) and without ( o ) 1 mM 3'-dA but not previously centrifuged on high-salt gradients were also analyzed. (Hiipakka and Liao, 1988)

The gradient-isolated receptor reanalyzed on a low salt gradient now sedimented at the same position (5-6S) as receptor from tissue incubated without 3'-dA (figure 6). Receptor from tissue incubated with 3'-dA but not subjected to high-salt sucrose gradient centrifugation sedimented at 8-9S. Gradient isolated receptors also eluted from DEAE- cellulose at salt concentrations similar to receptor from tissue incubated without 3'-dA (figure 5). Gradient-isolated receptor also had a DNA-cellulose-binding capacity similar to cytosolic receptor from tissue incubated without 3'-dA (69 vs 67 % of applied radioactivity bound), while only 7% of the cytosolic receptor from 3'-dA-treated tissue bound to DNA-cellulose. Since it appears that some components were removed during gradient centrifugation, we attempted to reverse this effect by adding back cytosol from tissue incubated with or without 3'-dA. However, we could not reverse the effect of gradient centrifugation by incubating gradient-isolated receptor with cytosol (data not shown).

## Discussion

The effect of 3'-dA on prostatic androgen receptors differed considerably from the effects of actinomycin D, another inhibitor of RNA synthesis. Incubation of prostatic tissue with actinomycin D increases chromatin-bound and decreases cytosolic levels of prostatic androgen receptors (Rossini et al.). Incubation of prostatic tissue with actinomycin D does not affect androgen-receptor binding to DNA-cellulose or alter the sedimentation characteristics of labeled androgen receptors as compared to receptor from tissue incubated without inhibitor (unpublished observations). The contrasting effects of actinomycin D and 3'-dA, both inhibitors of RNA synthesis, indicates that 3'-dA does not alter prostatic androgen receptor by inhibition of RNA synthesis. 3'-dA, at concentrations inhibiting androgen receptor transformation, can have little or no effect on the synthesis of heterogeneous nuclear RNA, but will still inhibit the polyadenylation of RNA (Penman et al.; Zeevi et al.). The effects of 3'-dA on the post-transcriptional processes of RNA polyadenylation and transport, therefore, may inhibit androgen receptor transformation.

We have proposed that steroid receptors interact with RNA in the cell as part of a recycling mechanism (Liao et al., 1972; Liao et al., 1973; Liang et al.; Liao et al., 1980). A hypothetical model of receptor recycling is given in figure 7. Interaction of receptor with RNA would facilitate removal of receptors from chromatin and may stabilize the RNA in contact with the receptor. Dissociation of RNA would be required before receptors could again interact with chromatin.

Although no direct proof is currently available establishing that steroid receptors bind to newly-synthesized, hormonally-regulated gene transcripts, evidence that RNA can bind to steroid receptors has accumulated over the past several years (see above). Autoradiographic studies have also suggested that in the rat ventral prostate, androgens stimulate intranuclear transport of RNA and migration of RNA through the nuclear envelope (Carmo-Fonseca). Evidence for estradiol stimulation of ribo-

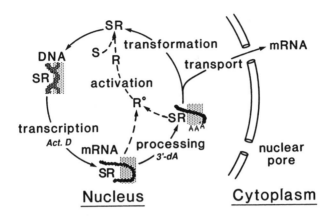

Fig. 7 A hypothetical model of intranuclear receptor cycling and its relationship to RNA processing and transport. A steroid-receptor complex (SR) binds to the transcriptional complex on the responsive gene and promotes RNA synthesis. If actinomycin D is present, SR accumulates at the site of transcription. In the absence of actinomycin D, SR can bind to the newly synthesized mRNA and is released from chromatin. By binding to RNA, SR may modulate processing, stabilization, and nucleo-cytoplasmic transport of RNA. Removal of RNA may allow SR to re-associate with chromatin. In the presence of 3'-dA, SR is maintained in a non-DNA-binding form. Some of the receptors may be inactivated to a form (R°) that does not bind steroid (S) (Rossini et al.; Mendel et al., 1986). R° may be reactivated by an energy-dependent process to the original form (R) that can bind steroid again . Dotted areas indicate that other proteins may be involved.

nucleoprotein (RNP) transport in rat uterine nuclei has also been presented (Thampan; 1985,1988). These effects may occur through a mechanism involving interaction of steroid-receptor complexes with RNA or RNP

The hormone-binding domain of a steroid receptor has been implicated in the masking of the DNA-binding domain of receptors (Godowski et al.; Hollenberg et al.; Kumar et al.). Direct interaction of these two regions or indirect physical constraints may maintain the DNA-binding domain in an inactive state. The steroid-binding domain of the glucocorticoid receptor also determines the formation of the 9S heteromeric receptor complex that does not bind to DNA (Pratt et al.). Therefore, this region may control the interaction of components, such as RNA and HSP 90 with the receptor and modulate DNA-binding activity. We could not reverse the transformation of androgen receptor (induced by high-salt sucrose gradient centrifugation) by addition of prostatic cytosol. Therefore, non-specific association of cytosolic components with the androgen receptor is not likely to be responsible for the untransformed state of the receptors. Since we were able to transform the receptor from 3'-dA-treated tissue by centrifugation of receptor on high-salt sucrose gradients or by addition of high concentrations (5 M) of urea, the 3'-dA effect was not due to an irreversible change , such as proteolytic removal of the receptor DNA-binding domain. High-salt sucrose gradient centrifugation and exposure to high concentrations of urea may transform androgen receptors by removal of a component(s) from the untransformed receptor.

We can not discount the possibility that 3'-dA inhibits androgen receptor-transformation by affecting processes besides RNA synthesis, polyadenylation and transport, since 3'-dA may be a competitive inhibitor of some processes involving adenosine. 3'-dA has been shown to inhibit the 2'-0-methylation of nuclear RNA (Glazer et al.) and the in vitro phosphorylation of non-histone chromosomal proteins (Legraverend et al.). These effects are not as well established as the effect of 3'-dA on RNA synthesis and polyadenylation. The inhibition of non-histone chromosomal protein phosphorylation may have some relevance to the effect of 3'-dA on androgen receptor. Steroid receptors are phosphoproteins and are phosphorylated after binding to nuclei (Logeat et al.; Pike et al.). However, the relevance of receptor phosphorylation to specific receptor functions is not known presently. Further study of the 3'-dA effect should be helpful in the understanding of the function and intracellular dynamics of steroid receptors and their regulation.

## Acknowledgments

We thank T. C. Popovich for her expert technical assistance. This work was supported by Grant DK09461 from the National Institutes of Health and Grant BC528 from American Cancer Society.

Abbreviations and definition used are: 3'-dA, 3'-deoxyadenosine; DMNT, $7\alpha,17\alpha$-dimethyl-19-nortestosterone. The term "receptor transformation" was used to indicate the process in which a steroid-receptor complex is converted from a form that does not bind to DNA (or chromatin) to a DNA (or chromatin)-binding form.

## References

Agutter PS, McCaldin B (1979) Biochem J 180: 371-378
Ali M, Vedeckis WV (1987) J Biol Chem 262: 6778-6784
Anderson EE, Tymoczko JL (1985) J Steroid Biochem 23: 299-306
Baez M, Sargan DR, Elbrecht A, Kulomaa MS, Zarucki-Schulz T, Tsai MJ,
    O'Malley BW (1987) J.Biol. Chem. 262: 6582-6588
Beach LR, Ross J (1978) J Biol Chem 253: 2628-2632
Burton K (1956) Biochem J 62: 315-323
Carmo-Fonseca M (1986) J Ultrastruct Res 94: 63-76

Chong MT, Lippman ME (1982) J Biol Chem 257: 2996-3002

Darnell JE, Philipson L, Wall R, Adesnik M (1971)Science 174: 507- 510

Diamond DJ, Goodman HM (1985) J Mol Biol 181: 41-62

Feldman M, Kallos J, Hollander VP (1981) J Biol Chem 256:1145-1148

Franceschi RT (1984) Proc Natl Acad Sci USA 81: 2337-2341

Glazer RI, Peale AL (1978) Biochem Biophys Res Commun 81: 521- 526

Godowski PJ, Rusconi S, Miesfeld R, Yamamoto KR (1987) Nature 325: 364-368

Hiipakka RA, Liao S (1988) J Biol Chem (In Press)

Hollenberg SM, Giguere V, Segui P, Evans RM (1987) Cell 49: 39-46

Horwitz KB, McGuire W L (1978) J Biol Chem 253: 6319-6322

Hutchens T W, Li CM, and Besch, P. K. (1987) Biochemistry 26, 5608-5616

Kletzein R F (1980) Biochem J 192: 753-759

Kumar V, Green S, Stack G, Berry M, Jin J, Chambon P (1987) Cell 51: 941-951

Legraverend M, Glazer RI (1978) Cancer Res 38: 1142-1146

Liang T, Liao S (1974) J Biol Chem 249: 4671-4678

Liao S, Fang S (1969) Vitam Horm 27: 17-90

Liao S, Tymoczko J L, Howell, DK, Lin AH, Shao TC, Liang T. (1972) Excerpta
    Medica Int Congr Ser 273: 404-407

Liao S, Liang T, Tymoczko JL (1973) Nature 241: 211-213

Liao S, Symthe S, Tymoczko JL, Rossini GP, Chen C, Hiipakka RA (1980) J Biol
    Chem 245: 5545-5551

Liao S, Witte D, Schilling K, Chang C (1984) J Steroid Biochem 20: 11-17

Lin S, Ohno S (1983) Biochim Biophys Acta 740: 264-270

Logeat F, Le Cunff M, Pamphile R, Milgrom E (1985) Biochem Biophys Res
    Commun 131:421-427

Martin RG, Ames BN (1961) J Biol Chem 256: 1372-1379

Mendel DB, Bodwell JE, Munck A (1986) Nature 324: 478-480

Mendel DB, Bodwell JE, Munck A (1987) J Biol Chem 262: 5644-5648

Mulder E, Vrij A A, Brinkmann AO ,Van Der Molen HJ, Parker, MG (1984)
    Biochim Biophys Acta 781: 121-129

Page MJ, Parker MG (1982) Mol Cell Endocrin 27: 343-355

Penman S, Rosbash M, Penman, M (1970) Proc Natl Acad Sci USA 67: 1878-
    1885

Pike JW, Sleator NM (1985) Biochem Biophys Res Commun 131:378-385

Plunkett W, Cohen SS (1975) Cancer Res 35: 1547-1554

Pratt WB, Jolly DJ, Pratt DV, Hollenberg SM, Giguere V, Cadepond FM,
    Schweizer-Groyer G, Catelli MG, Evans RM, Baulieu EE (1988) J Biol Chem
    263: 267-273

Riehl RM, Sullivan W P, Vroman BT, Bauer VJ, Pearson, GR, Toft DO (1985)
    Biochemistry 24: 6586-6591

Rossini GP, Liao S (1982) Biochem J 208: 383-392

Rossini GP (1985) J Steroid Biochem 22: 47-56

Rowley DR, Premont RT, Johnson MP, Young CYF, Tindall DJ (1986)
    Biochemistry 25: 6988-6995

Sanchez ER, Meshinchi S, Tienrungroj W, Schlesinger MJ, Toft DO, Pratt WB

(1987) J Biol Chem 262: 6986-6991

Schmidt TJ,  Litwack G (1982) Physiol Rev 62: 1131-1192

Schoenberg DR,  Clark JH (1979) J Biol Chem 254: 8270-8275

Shapiro D J,  Brock ML (1985) in Biochemical Actions of Hormones ( Litwack G ed) Vol 12, pp 139-172  Academic Press, New York

Shapiro DJ, Blume JE,  Nielsen DA (1987) BioEssays 6: 221-226

Smith AC, Elsasser MS,  Harmon JM (1986) J Biol Chem 261: 13285-13292

Thampan RV (1985) J Biol Chem 260:5420-5426

Thampan RV (1988) Biochemistry 27:5019-5026

Traish AM, Muller RE,  Wotiz HH (1985) J Steroid Biochem 22: 601-609

Tymoczko JL,  Phillips MM (1983) Endocrinology 112: 142-149

Yamamoto KR (1985) Annu Rev Genet 19: 209-252

Zeevi M, Nevins JR,  Darnell JE Jr (1982) Mol Cell Biol 2: 517-525

DISCUSSION OF THE PAPER PRESENTED BY S. LIAO

CIDLOWSKI: Does the recycling event occur in the cytoplasm or the nucleus?

LIAO: Using polyclonal antibodies generated against protein products produced from AR-cDNA, Drs. Michael Press and Tony Antakly, with whom we are collaborating, have found that androgen receptors are mainly localized inside the cell nuclei in human prostate (normal and cancerous) and liver. Therefore, we believe that the major recycling event I have described occurs inside the cell nucleus of target organs.

CIDLOWSKI: Can you speculate on the mechanism of reactivation of receptor to the steroid binding form?

LIAO: The work from Dr. Munck's and our laboratories in the 1960's showed that both glucocorticoid receptors and androgen receptors required an energy dependent process for the activation of receptors that does not bind steroids to the steroid binding form. Some investigators have speculated that protein phos-phorylation may be involved but no one knows exactly how the receptor reactivation is carried out in the intact cells. This reactivation is, of course, different from transformation (often called activation) which is required for conversion of steroid-receptor complexes to DNA/chromatin binding form. The steps involved in the transformation have not been clearly identified.

TATA: Does the number of AR molecules decrease upon castration?

LIAO: Our observations suggest that the number of androgen receptor molecules per cell in the rat ventral prostate do not decrease significantly within two days after castration. However, androgens may autoregulate the level of androgen receptor mRNA in the prostate. A dot hybridization analysis using an androgen receptor cDNA probe showed that the androgen receptor mRNA level per unit of poly(A+)RNA increased to 140% of normal two days after castration.

DISCUSSANTS:  J. CIDLOWSKI, S. LIAO, J.R. TATA

# ANDROGEN ACTION IN RAT LIVER: CHANGES IN ANDROGEN SENSITIVITY DURING MATURATION AND AGING

**B. CHATTERJEE, W.F. DEMYAN, W. GALLWITZ, J.M. KIM, M.A. MANCINI, D.H. OH, C.S. SONG** and **A.K. ROY**, University of Texas Health Science Center, Departments of Cellular & Structural Biology and Obstetrics & Gynecology, San Antonio, TX

## INTRODUCTION

The liver plays both direct and indirect roles in reproduction. In the oviparous female, the liver is the source of most of the yolk proteins and in mammals, it is also responsible for maintaining the sexually dimorphic pattern of steroid metabolism (Roy & Chatterjee, 1983; Gustafsson et al, 1983). A large number of studies investigating the effect of estrogens on the hepatic synthesis of vitellogenins have provided considerable understanding of the mechanism of estrogen action in the liver (Tata et al, 1987). The paucity of suitable model systems, however, has impeded similar scrutinization of androgen action on hepatocytes. Following our initial discovery of $\alpha$2u globulin and its androgen-dependent synthesis in the rat liver (Roy & Neuhaus, 1967), this protein has served as an important model for studying androgenic regulation of hepatic gene expression. In addition, studies with the analogous mouse urinary protein (MUP) have also been used to explore the mechanism of androgen-mediated gene expression in hepatocytes (Hastie et al, 1979). Subsequently, we have identified two other androgen-dependent hepatic proteins, i.e. senescence marker protein 1 (SMP-1) and senescence marker protein 2 (SMP-2) and have developed the SMP-2 system as a model to study androgenic repression of gene expression (Chatterjee et al, 1981; Chatterjee et al, 1987). Similar to $\alpha$2u globulin, SMP-1 ($M_r$ 34,000) is an androgen-inducible protein; the hepatic expression of SMP-2 ($M_r$ 31,000), however, is androgen repressible (Chatterjee et al, 1984). A male-specific mouse liver protein (sex-limited protein, Slp), possibly the mouse analog of SMP-1, has also been identified (Hemenway & Robins, 1987). A recent report shows that a proviral long terminal repeat element, integrated at around 2 kb upstream of the Slp gene, confers androgen sensitivity to this protein (Stavenhagen & Robins, 1988). In addition to these non-enzymatic secretory and intra-cellular proteins, synthesis of a number of steroid and drug metabolizing enzymes, such as carbonic anhydrase, glutathione S-transferase and histidase, are also regulated by androgens (Roy & Chatterjee, 1983). This article will review our current state of knowledge of androgen action in the rat liver as gleaned through studies on $\alpha$2u globulin and SMP-2 gene expression.

Although $\alpha$2u globulin, and more recently SMP-2, have served as important models for studying androgen-mediated gene expression, the biological functions of these two proteins and their possible roles in the reproductive process are just beginning to be understood. $\alpha$2u globulin has a strong structural homology with a number of lipid carrier proteins (Pervaiz & Brew, 1987), and has been localized in association with lipid droplets in several pheromone producing glands (e.g. preputial, perianal and meibomian), indicating its likely role in the transport of reproductively relevant pheromonal lipids (Mancini et al, 1989). Such a notion is further suggested by non-covalent binding of $\alpha 2u$ globulin to aliphatic hydrocarbons (Lock et al, 1987).

In the case of SMP-2, preliminary studies involving immunostaining of rat hepatocytes show that it is localized on the cell membrane. Furthermore, the nucleotide sequence determination up to 2 kb of the upstream SMP-2 gene reveals the presence of several metal regulatory consensus sequences TGC $\binom{A}{G}$C$\binom{C}{A}$C, similar to those found in the metallothionein gene. The element CTGGGA, whch is the consensus sequence for the acute-phase signal, also occurs several times in the regulatory region of this gene. These results suggest a possible metal transporting membrane function for SMP-2.

### Structural Features of $\alpha 2u$ Globulin, SMP-2 and their Corresponding Genes

$\alpha 2u$ globulin in the rat and the analogous mouse urinary protein (MUP) are coded by a multigene family comprised of about 30 highly conserved ( 90%) gene copies per haploid genome (Hastie et al, 1979; Kurtz, 1981). At least ten of these genes are expressed in the liver. The rat genes are clustered on chromosome 5, whereas the mouse genes, the organization of which has been studied more extensively, are located on chromosome 4. The mouse genes are arranged as 45 kb units of imperfect palindromes containing two MUP gene copies connected in a head-to-head fashion. Some of these palindromic units (45 kbp) are in direct tandem repeats, while others are in inverted orientations with respect to the adjacent units (Bishop et al, 1985). The complete area of the MUP locus covers about 600-700 kbp of chromosome 4. Comparative sequence analysis of several genomic clones of $\alpha 2u$ globulin shows an A-rich region similar to what is found in other androgen-regulated genes such as the $C_1$, $C_2$ and $C_3$ subunits of the prostatic steroid binding protein. For the $\alpha 2u$ globulin gene, a similar region is located around 378 bp upstream of the transcription start site (Windzickx et al, 1987). Each member of the $\alpha 2u$ globulin gene family contains seven exons and six introns, and the processed mRNAs code for proteins of 162 amino acid residues (Dolan et al, 1982).

The cDNA sequence for androgen-repressible SMP-2 shows an open reading frame for 282 amino acids containing, at the carboxy terminus, eleven consecutive hydrophobic residues (Chatterjee et al, 1987). The very last amino acid at the c-terminal, however, is acidic in nature.This finding further supports the results of immunocytochemical experiments and indicates membrane localization of this protein. The rat genome seems to have a very low copy number for the SMP-2 sequence, as digestion of the genomic DNA with any of several restriction enzymes followed by electrophoresis and Southern hybridization of the blotted gel produce a simple hybridization pattern.

### Androgenic Regulation of $\alpha 2u$ Globulin and SMP-2

The hepatic synthesis of $\alpha 2u$ globulin begins in the male rat during puberty ( 40 days), reaches a peak level at around 100 days of age and thereafter gradually declines to a non-detectable level beyond 750 to 800 days of age (Roy et al, 1983). Mature female rats, which normally do not synthesize $\alpha 2u$ globulin in the liver, can be induced to synthesize this protein following ovariectomy and subsequent androgen treatment. Several other hormones (i.e. growth hormone, insulin, thyroxine and glucocorticoids), in addition to androgen, influence the expression of this protein (Roy et al, 1983a).

The secretory patterns of growth hormone in male and female animals are different and the differential regulation of certain sexually dimorphic hepatic enzymes are explainable on the basis of the sex-specific secretory patterns of growth hormone (Gustafsson et al, 1983). For α2u globulin synthesis also, the influence of androgen was previously suggested to be indirect, acting through the hypothalamic-pituitary system (Norstedt & Palmiter, 1984). Inhibition of α2u globulin synthesis after administration of excess growth hormone through osmotic minipumps (mimicking the feminine pattern) was presented as the supporting evidence for such an indirect influence. Recently, however, through an in vitro liver perfusion system, we have clearly demonstrated that the androgen does indeed act directly and rapidly on the liver to stimulate α2u globulin synthesis (Fig. 1) (Murty et al, 1987a). The presence of the androgen receptor mRNA in the mouse liver, as seen in the Northern blot analysis, provides further credence to a direct androgen action in the hepatic tissue (Lubahn et al, 1988).

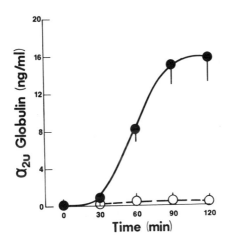

**Figure 1:** Cumulative levels of α2u globulin in the perfusates of ovariectomized female rat livers with and without in vitro androgen supplementation. Ovariectomized rats were pretreated for 6 days with daily injections of DHT (50 ug/100 g). Livers were removed 24 h after the final androgen treatment and perfused with a perfusion medium containing the blood from castrated rabbits. DHT (300 ug; ●—●) or vehicle (5 ul ethanol; O--O) was added to the perfusion fluid at 0 min. Samples were taken at 30 min intervals, and α2u globulin concentrations were determined by RIA. (From Murty et al., 1987a)

The hepatic expression of the SMP-2 gene appears to be under androgenic repression. The female liver has a much higher steady-state level of the SMP-2 RNA than does the liver of the adult male rat. Androgen treatment of ovariectomized female rats almost completely abolishes SMP-2 mRNA in the liver. A high level of the SMP-2 mRNA is also found in the liver of testicular feminized male rats (Fig. 2), (Chatterjee et al, 1987). Thus, whenever the liver attains an androgen-insensitive state, the expression of the SMP-2 gene is enhanced.

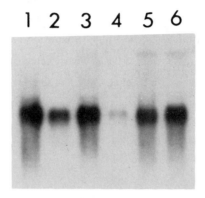

**Figure 2**: Northern blot analysis for the SMP-2 mRNA within poly(A)[+] hepatic RNAs from: 1, adult female; 2, adult male; 3, ovariectomized female; 4, ovariectomized female injected daily for 2 weeks with 5α-dihydro-testosterone (50 ug/100 g of body weight); 5, testicular feminized male; 6, testicular feminized male treated for 2 weeks with estradiol-17β(10 ug/100 g). (From Chatterjee et al., 1987)

---

### Cytoplasmic Androgen Binding Protein and Changes in Hepatic Androgen Sensitivity

As mentioned earlier, the temporal expression of α2u globulin and SMP-2 follows an opposite, yet complementary pattern. The prepubertal male rat has a low expression of α2u globulin, but a high expression of SMP-2. After puberty, the synthesis of α2u globulin and its mRNA rapidly increases, with concomitant inhibition of SMP-2 gene expression. The senescent rat liver, however, shows a derepression of the SMP-2 gene and repression of the gene for α2u globulin. During the course of maturation and aging, the hepatic tissue of the rat passes through three distinct phases, i.e. prepubertal androgen-insensitivity, androgen responsiveness during adulthood, and androgen insensitivity during senescence. Investigation of the biochemical basis of such temporal changes in androgen sensitivity has led us to the identification and characterization of a cytoplasmic androgen binding (CAB) protein in the rat liver, the presence and absence of which, during different phases of the life, correlate with androgen sensitivity of the tissue. This low-capacity ( 50 fmole/mg total protein) high affinity ($K_d = 10^{-8}$ M) binding protein has a high degree of androgen specificity (Table I). The correlation between the presence of the androgen binder and the ability of the liver to synthesize α2u globulin is presented in Table II.

## TABLE I

### Specific Binding of [$^3$H]R-1881 to Adult ( 100-day-old) Male Liver Cytosol and Competition by Nonandrogenic Steroids

| competing steroid | % displacement | competing steroid | % displacement |
|---|---|---|---|
| R-1881 | 100 | progesterone | 5.8 |
| triamcinolone acetonide | 6.2 | estradiol-17β | 31.1 |

Binding assays were performed in the presence of 58 nM [$^3$H]R-1881 and a 500-fold molar excess of unlabeled competing steroids. In the absence of any unlabeled competing steroids, 52.0 fmole of R-1881 was specifically bound to 1 mg of cytosolic protein (100% binding). (From Sarkar et al., 1987)

## TABLE II

### Correlation between Age- and Sex-Specific Binding of [$^3$H]R-1881 to Liver Cytosol and Level of Hepatic α2u Globulin

| liver cytosol | specific R-1881 binding (fmol/mg of protein) | cytosolic α2u globulin (ng/mg of protein) |
|---|---|---|
| adult male (100 days) | 52.98 | 240.00 |
| adult female (100 days) | 6.53 | 0.85 |
| immature female (30 days) | 5.14 | 0.41 |
| old male | 7.99 | 2.66 |

Values are means of three animals for each age group. (From Sarkar et al., 1987)

Photoaffinity labeling of the liver cytosol with tritiated methyltrienolone (R-1881) and SDS-polyacrylamide slab gel electrophoresis of the labeled proteins have shown that CAB has a molecular mass ($M_r$) of 31,000 (Sarkar et al, 1987). In an effort to understand the relationship of CAB with the rat androgen receptor ($M_r$ 76 kilodalton), we are presently considering the following three possibilities: (1) CAB as a proteolytic fragment of the androgen receptor, (2) CAB as the product of alternate splicing of the androgen receptor mRNA, and (3) CAB as a protein different from the androgen receptor, but belonging to the same family of androgen-binding proteins. We have purified the CAB protein and raised the monospecific antiserum. Work is in progress in our laboratory to establish the amino acid sequence of CAB and to identify the cDNA clone corresponding to this protein. Use of the CAB-specific polyclonal antibody in the Western-blot analysis has shown that the triphasic changes in the hepatic concentration of CAB during maturation and aging, as observed through steroid binding assays, are due to changes in the actual concentration of the CAB protein (Fig. 3) (Demyan et al, 1989).

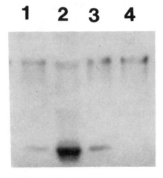

**Figure 3**: Immunoblot for the CAB protein within the liver cytosol of rats of different ages and sex. Lanes are marked as follows: 1, prepubertal male (30 day old); 2, young adult male (100 day old); 3, senescent males, (850 day old); 4, adult female (100 day old). (From Demyan et al., 1989)

It is intriguing to note that, although a stringent positive correlation exists between hepatic androgen sensitivity and the presence of the 31 kDa CAB in the liver cytosol, the androgen-insensitive state of the liver correlates with the presence of a 29 kDa weak androgen-binding protein (Sarkar et al, 1987). The ligand-protein association of the 29 kDa androgen binder does not survive sucrose density gradient centrifugation and thus went undetected in our earlier studies. The polyclonal antibody raised against the 31 kDa androgen binder does not react with the 29 kDa binding component and the molecular relationship between these two binding moieties is presently unclear. Elucidation of this relationship may provide important clues to the mechanism of programed changes in androgen sensitivity during aging.

### Role of Nuclear Matrix and a Trans-acting Regulatory Protein in α2u Globulin Gene Expression

Transcriptional activation of a gene or gene family is known to be controlled by several independent events. Decompaction of the chromatin structure containing the gene domain seems to be one of the earliest steps in its eventual expression. Earlier studies in our laboratory have shown that the chromatin regions containing the α2u globulin gene first attain preferential

DNaseI sensitivity around 22 days of age, and this "open" conformation of the $\alpha$2u globulin gene domain is maintained for the rest of the animal's life (Roy et al, 1983a). However, transcription of the $\alpha$2u globulin gene family does not begin until about 40 days of age and almost ceases beyond 750 days. A similar temporal pattern of changes in DNase I sensitivity of the gene for the sex-limited protein (Slp) in the mouse has also been reported (Hemenway & Robins, 1987).

Subsequent to chromatin decompaction, a regulatory event in the transcriptional activation seems to be the association of the DNase I sensitive gene domains with the nuclear matrix. Growing bodies of evidence support the concept that the DNA strands in the eukaryotic nucleus are organized around a protein framework called the nuclear matrix. Such a protein framework has also been referred to as the nuclear scaffold or nuclear skeleton. Upon removal of histones, the extended DNA loops are still found to be firmly attached to this matrix network. Although the precise functional role of the nuclear matrix is still speculative, a number of actively transcribed genes have been shown to be preferentially associated with the nuclear matrix, and steroid hormone receptors, as well as various regulatory proteins, such as DNA and RNA polymerases, are known to be localized in the matrix framework (Barrak & Coffey, 1983). Thus, matrix association of certain gene domains seems to be an important regulatory step in their transcription and eventual expression. Our studies show that association and dissociation of the $\alpha$2u globulin gene domain with and from the nuclear matrix are related to the transcriptional activity of this gene during maturation and aging (Murty et al, 1987b).

The age- and sex-specific changes in the transcription of the $\alpha$2u globulin gene are reflected in the concentrations of the $\alpha$2u globulin mRNA within the total poly(A)-containing hepatic RNA of male rats (Roy et al, 1983b). The livers of prepubertal (28-day-old) male, 29-month-old male and female rats of all ages do not contain any significant concentrations of the $\alpha$2u globulin mRNA, as compared to young adult (100-day-old) males. That such changes in the steady-state levels of the mRNA are due to differences in the transcriptional rates of this gene is shown in the experimental results presented in Table III. The liver of the prepubertal male (30 days), which contains no detectable $\alpha$2u globulin mRNA, also showed an almost undetectable rate of $\alpha$2u globulin gene transcription. The high level of $\alpha$2u globulin gene transcription in the young adult male (92 days) is reduced by about 56% in the 14-month-old male rat and is reduced to about 6% by the 25th month of age.

### TABLE III

**Transcriptional Rate of the $\alpha$2u Globulin Gene During Aging**

| Age of the animal (days) | $\alpha$2u globulin gene transcription (sp. hybridizable dpm per 6 x 10$^6$ dpm of total RNA transcripts) | Transcriptonal activity compared to the young adult (%) |
| --- | --- | --- |
| 30 | 30 (36, 24, 31) | 2.5 |
| 92 | 1200 (1192, 1222, 1188) | 100 |
| 425 | 683 (663, 670, 718) | 56 |
| 760 | 76 (63, 67, 100) | 6.3 |

(From Murty et al, 1987b)

The α2u globulin and albumin DNA sequences were quantitated in the total nuclear DNA, matrix associated DNA and, the DNA fragments that are released by EcoRI digestion of the salt and detergent extracted liver nuclei (Fig. 4). Enrichment of the α2u globulin DNA sequence in the nuclear matrix can be found only in the hepatocytes of young adults (150 days old), where the gene is maximally transcribed. DNA extracted from the released fraction or the DNA bound to the matrix showed no selective enrichment when the 28-month-old male rat liver or the kidney was examined. The α2u globulin gene is silent in the kidneys, and the liver of the 28-day-old (prepubertal) male rat does not synthesize this protein. In the case of the 900-day-old male rat, however, where the α2u globulin gene almost ceases to be transcribed, the pattern of distribution of the α2u globulin gene is not exactly same as that of the prepubertal male rat. As compared to the 28-day-old male, we have found that a disproportionately higher amount of α2u globulin DNA is present in the released fraction. Presently, we do not have any satisfactory explanation for this observation.

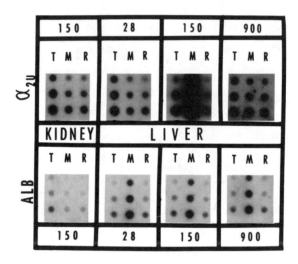

**Figure 4**: Dot-blot analysis of α2u globulin and albumin genes in the liver and kidney nuclear preparations. DNA was extracted from total nuclei (T), salt and EcoRI-resistant nuclear matrix (M), and the released fraction obtained after digestion of the salt-resistant structure with EcoRI (R). 0.5, 1.0 and 2.0 ug of each preparation of DNA were spotted on the nitrocellulose paper and hybridization was performed with either α2u globulin (a2u) or albumin (ALB) cDNA probes. Numbers at the top and bottom represent the age of the donor animals in days. The first two frames on the left are dot-blots of the kidney DNA and the remaining six frames are liver DNA. (From Murty et al., 1987b)

The albumin gene sequence, used as an age-invariant control, was enriched in the matrix fraction of the liver DNA at all ages; as expected, no matrix association for this gene is seen in the kidney. It should be mentioned that only one copy of the albumin gene is present per haploid genome, in contrast to

25-30 copies for the α2u globulin gene. The difference in the autoradiographic intensity of these two matrix-associated genes is a reflection of this variable.

The lack of a temporal correlation between DNase I sensitivity and matrix association of the α2u globulin gene indicates that these two events are independently regulated. Studies with the regulation of ovalbumin gene expression have also indicated that a segment of this gene domain, which is DNase I sensitive but not expressed, is not bound to the nuclear matrix (Ciezek et al, 1983). Furthermore, inactivation of ovalbumin gene transcription with actinomycin D does not result in the release of this gene from the nuclear matrix, suggesting that the process of transcription itself is not the determinant event for matrix association of this gene (O'Malley et al, 1983).

## Nutritional Modulation of Age-Dependent Changes in Androgen Responsiveness of the Rat Liver

Food restriction has to date remained the only reproducible means for prolonging mean and maximum life span of rodents and several other animal species. Although the mechanism of this important phenomenon is not clear, initial theoretical analysis by Sacher (1977) has led to the conclusion that caloric restriction increases longevity by retarding the aging process. We have explored the possible retardation of the age-dependent loss of hepatic androgen sensitivity in calorie-restricted male rats.

A comparative analysis of the hepatic mRNA for α2u globulin in male rats of 6 to 27 months of age, which are fed ad libitum, shows a marked reduction of this mRNA at 24 months and an almost 20-fold reduction (5% of the 6-month level) at 27 months. On the other hand, in animals subjected to a 40% reduction in food consumption since 42 days of age, 55% of the steady-state level of mRNA was maintained at 27 months, as compared to the 6 months' level. Quantitatively, this is equivalent to about 9 months of delay in the progression of the aging process within the period of 27 months. A decreased level of the androgen repressible SMP-2 mRNA in calorie-restricted animals provides strong supportive evidence for the molecular basis of calorie-restriction in retardation of the aging process (Chatterjee et al, 1989). Results showing a delay in the age-dependent loss of the CAB protein and α2u globulin and a rise in SMP-2 are presented in figure 5. Thus it appears that food restriction is highly effective in retarding the age-dependent loss in hepatic androgen sensitivity.

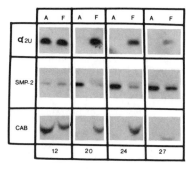

**Figure 5:** Western blots of immunoreactive α2u globulin, SMP-2 and CAB in the liver cytosols obtained from rats of progressively increasing age

maintained on either the ad libitum (AL) or restricted (FR) diet. Same amount of total protein (30 ug) was applied to each lane. (From Chatterjee et al, 1989).

## SUMMARY AND CONCLUSIONS

We have utilized α2u globulin gene expression in the rat liver as a model to explore changes in hepatic androgen sensitivity during maturation and aging. Our results show that the age-dependent changes in hepatic androgen sensitivity can be clearly divided into three different stages, i.e. prepubertal androgen insensitivity, androgen responsiveness during adulthood, and an androgen insensitive state during senescence. These three stages not only dictate the temporal differences in the expression of the α2u globulin gene, but they also reflect the basis of changes in the expression of other androgen-sensitive genes such as SMP-1 and SMP-2. Age-dependent changes in hepatic androgen sensitivity are mediated through multiple regulatory processes involving developmental, hormonal and nutritional factors. Preferential DNase I sensitivity, signifying opening of the α2u globulin gene domain during 22-24 days of life, may be developmentally programmed. Matrix association and transcriptional activation during puberty ( 40 days) are dependent on androgenic induction. Androgenic induction during puberty may be dependent on the developmentally programmed appearance of the cytoplasmic androgen-binding protein. The coordinated effects of the androgen and the androgen binding protein are reflected in the pubertal appearance of the sequence-specific trans-acting binding factor which, in turn, is responsible for transcriptional activation of the gene. These regulatory events seem to be reversed during senescence. However, nutritional studies show that the time of onset of such reversal can be markedly delayed by caloric restriction. The complexity of the regulatory network and its precise and timely coordination are hallmarks of the manifestation of a physiological event critical for efficient reproductive function.

This work is supported by grants AG-03527 and DK-14744.

## REFERENCES

Bishop JO, Selman GG, Hickman J, Black L, Saunders RDP and Clark AG (1985) The 45-kb unit of major urinary protein gene organization is a gigantic imperfect palindrome. Mol Cell Biol 7:1591-1600.

Barrack ER and Coffey DS (1983) Hormone receptors and the nuclear matix. In: Gene Regulation by Steroid Hormones II (AK Roy & JH Clark, eds) pp. 239-266, Springer-Verlag, New York.

Chatterjee B, Fernandes G, Yu BP, Song CS, Kim JM and Roy AK (1989) Calorie restriction delays age-dependent loss in androgen responsiveness of the rat liver. FASEB J. (in press)

Chatterjee B, Nath T and Roy AK (1981) Differential regulation of the messenger RNA for three major senescence marker proteins in the male rat liver. J Biol Chem 256:5939-5941.

Chatterjee B, Ozbilen O, Majumdar D and Roy AK (1987) Molecular cloning and characterization of the cDNA for androgen repressible rat liver protein SMP-2. J Biol Chem 262:822-825.

Chatterjee B, Murty CVR, and Roy AK (1984) Androgenic repression of the messenger RNA for a 26.3 kilodalton hepatic protein in the rat. FEBS Lett. 170:114-117.

Ciezek EM, Tsai MJ and O'Malley BW (1983) Actively transcribed genes are associated with the nuclear matrix. Nature 306: 607-609.

Demyan WF, Sarkar FH, Murty CVR and Roy AK (1989) Purification and immunochemcal characterization of the cytoplasmic androgen binding protein of the rat liver. Biochemistry (in press)

Dolan KP, Unterman R, McLaughlin M, Nakhasi HL, Lynch KR and Feigelson P (1982) The structure and expression of very closely related members of the α2u globulin gene family. J Biol Chem 257:13527-13534.

Gustafsson JA, Mode A, Norstedt G and Skett P (1983) Sex steroid induced changes in hepatic enzymes. Annual Review of Physiology 45:51-60.

Hastie ND, Held WA and Toole JJ (1979) Multiple genes coding for the androgen-regulated major urinary proteins of the mouse. Cell 17:449-456.

Hemenway C and Robins DM (1987) DNase I-hypersensitive sites associated with expression and hormonal regulation of mouse C4 and Slp genes. Proc Natl Acad Sci USA 84:4816-4820.

Kurtz DT (1981) Rat alpha 2u globuin is encoded by a multigene family. J Mol Appl Genet 1:29-38.

Lock EA, Charbonneau M, Strasser J, Swenberg JA and Bus JA (1987) 2,2,4-trimethylpentane-induced nephrotoxicity. The reversible binding of a TMP metabolite to a renal protein fraction containing alpha 2u globulin. Toxicol Appl Pharm 91:182-192.

Lubahn DB, Joseph DR, Sullivan PM, Willard HF, French FS and Wilson EM (1988) Cloning of human androgen receptor complementary DNA and localization to the X chromosome. Science 240:327-330.

Mancini MA, Majumdar D, Chatterjee B and Roy AK (1989) α2u globulin in modified sebaceous galnds with pheromonal functions: Localization of the protein and its mRNA in preputial, meibomian and perianal glands. J Histochem Cytochem (in press)

Murty CVR, Rao KVS and Roy AK (1987a) Rapid androgenic stimulation of α2u globulin synthesis in the perfused rat liver. Endocrinology 121:1814-1818.

Murty CVR, Mancini MA, Chatterjee B and Roy AK (1987b) Changes in transcriptional activity and matrix association of α2u globulin gene family in the rat liver during maturation and aging. Biochim Biophys Acta 949: 27-34.

Norstedt G and Palmiter R (1984) Secretory rhythm of growth hormone regulates sexual differentiation of mouse liver. Cell 38:805-812.

O'Malley BW, Tsai MJ and Schrader WT (1983) In: Steroid Hormone Receptors: Structure and Function (H Eriksson & JA Gustafsson, eds.) pp. 307-328, Elsevier, Amsterdam.

Pervaiz S and Brew K (1987) Homology, structure function correlations between $\alpha$1-acid glycoprotein and serum retinol-binding protein and its relatives. FASEB J 1:209-214.

Roy AK and Chatterjee B (1983) Sexual dimorphism in liver. In: Annual Review of Physiology, Vol.45.(RM Berne, ed.), Annual Review, Palo Alto, California, pp. 37-50.

Roy AK and Neuhaus OW (1967) Androgenic control of a sex-dependent protein in the rat. Nature 214:618-620.

Roy AK, Chatterjee B, Demyan WF, Milin BS, Motwani NM, Nath T and Schiop MJ (1983a) Hormone and age-dependent regulation of $\alpha$2u globulin gene expression. In: Recent Progress in Hormone Research (RO Greep, ed.) Academic Press, pp. 426-461.

Roy AK, Nath TS, Motwani NM and Chatterjee B (1983b) Age and androgen dependent regulation of polymorphic forms of $\alpha$2u globulin. J Biol Chem 258:10123-10127.

Sacher GA (1977) Life table modification and life prolongation. In: Handbook of the Biology of Aging (CE Finch and L Hayflick, eds.) Van Nostrand Reinhold, pp. 582-638.

Sarkar FH, Sarkar PK, Watson S, Poulik MD and Roy AK (1987) Cytoplasmic androgen binding protein of the rat liver: Molecular characterization after photoaffinity labeling and functional correlation with $\alpha$2u globulin synthesis during maturation and aging. Biochemistry 26:3965-3970.

Stavenhagen JB and Robins DM (1988) An ancient provirus has imposed androgen regulation on the adjacent mouse sex-limited protein gene. Cell 55:247-254.

Tata JR, Ng WC, Perlan AG, Wolffe AP (1987) Activation and regulation of the vitellogenin gene family. In: Gene Regulation by Steroid Hormones III (AK Roy & JH Clark, eds.) Springer-Verlag, pp. 205-233.

Windzickx D, VanDijck P, Dirckx G, Valkaert G, Rombauts W, Heyns W and Verhoeven G (1987) Comparison of the 5' upstream putative regulatory sequences of three members of the $\alpha$2u globulin gene family. Eur J Biochem 165: 521-529.

DISCUSSION OF THE PAPER PRESENTED BY A.K. ROY

KORACH: Do you know the pattern of expression of the binding protein and is it similar to the alpha 2u expression? Also, have you checked cross reactivity of your binding protein antibody with prostate androgen receptor?

ROY: The answer to your first question is yes, - the age-dependent expression of $\alpha_{2u}$ globulin parallels the appearance of the CAB protein. With respect to cross reactivity, we have tried one of Dr. Liao's antibodies and it does not react with the CAB protein.

CIDLOWSKI: Arun, in the interesting dietary restriction studies you showed was there a change in the diets composition?

ROY: The vitamin and mineral intakes of both restricted and control animals are maintained the same while the calorie content of proteins, fats and carbohydrates are equally reduced by 40%.

JUMP: In your studies on the expression of alpha 2u globulin in the perivenous area, you reported data on immunohistochemistry. Do you have similar data on the in situ hybridization? Is alpha globulin protein migrating from the perivenous to periportal region?

ROY: Yes, both immunocytochemical staining and in situ hybridization with cloned RNA shows the same cellular localization. We, therefore, can safely conclude that distribution pattern reflects selective expression of the $\alpha_{2u}$ globulin gene.

JUMP: Do you have or are you aware of androgen effects on gap junction structure in liver?

ROY: Presently, I do not have any idea about the role of androgen on gap junction in the liver.

HARDIN: Do liver cubes or minces respond to androgens and induce alpha 2u globulin?

ROY: Yes, the liver cubes show limited androgen sensitivity with respect to $\alpha_{2u}$ globulin synthesis.

TATA: Do you know why estrogen renders the liver cells refractory to respond to androgen? Also, with regard to your failure to induce alpha 2u in primary culture, did you allow the cells to reform cell-cell interaction pattern before treatment with androgen?

ROY: My guess will be that estrogen causes certain biochemical changes which prevent expression of the CAB protein. Many years ago we performed a parabiosis experiment where one of the partner was pretreated with estradiol and the other was the control. Result of this experiment indicated that the estrogenic refractoriness is not due to any humoral factor. In answer to your second question - I do not think we have carefully looked into that.

DISCUSSANTS: K. KORACH, A.K. ROY, J. CIDLOWSKI, D. JUMP, J. HARDIN AND J. TATA.

# INHIBITION OF CELL PROLIFERATION BY A NUCLEAR TYPE
II LIGAND: METHYL P-HYDROXYPHENYLACTATE

Barry M. Markaverich*, Brian S. Middleditch** and James Clark*
*Dept. of Cell Biology, Baylor College of Medicine, Houston, TX 77030
**Department of Biochemical and Biophysical Sciences University of
Houston, Houston, TX 77004

## INTRODUCTION

Previous studies from our laboratory demonstrated that a variety of
rat tissues contain endogenous ligands for nuclear type II sites.
Since type II sites bind to [³H]estradiol with a relatively low
affinity (Markaverich and Clark, 1979) and have a higher affinity for
the endogenous ligands (Markaverich et al., 1983; 1984), we proposed
that these compounds may regulate cell growth by their association
with unclear type II binding sites. This hypothesis was supported by
a number of studies which suggest that nuclear type II sites may be
involved in cell growth regulation by estrogenic hormones (Markaverich
et al., 1981, Clark et al., 1982). Moreover, our earlier experiments
demonstrated that normal tissues contained two type II ligands (α and
β) which we separated on Sephadex LH-20 (2). In contrast, malignant
tissues contained only the α-peak component, suggesting that a
deficinecy in the β-peak material was characteristic of tumor tissue
(Markaverich et al., 1984). Our preliminary experiments also
demonstrated that partially purified pareparations of the β-peak
component inhibited the growth of MCF-7 human breast cancer cells in
vitro, whereas this inhibition was not observed with the α-peak
component.

## IDENTIFICATION OF TYPE II LIGANDS

We were able to isolate sufficient amounts of the α- and β-peak
components from fetal bovine serum for complete qualitative
analysis. We have data which demonstrate that the α-peak component is
hydroxyphenylactate (HPLA) and the β-peak component is methyl para-
hydroxyphenylactate (MeHPLA). These are summarized below and
described in detail in Markaverich et al., 1988.

1. The isolated and authentic compounds have the same retention
   behavior during chromatography on Sephadex LH-20.
2. The isolated and authentic compounds have the same retention
   behavior during HPLC on Ultrasphere-Octyl.
3. The isolated and authentic compounds have the same gas
   chromatogrphic retention index values.
4. Trimethylsilyl derivatives of the isolated and authentic compounds
   have the same mass spectra.
5. The isolated and authentic compounds have the same ultraviolet
   absorption spectra.
6. The isolated and authentic compounds have the same specific
   binding activities to the type II site.
7. Neither the isolated nor the authentic compounds bind to the
   (conventional) rat uterine estrogen receptor.

In all respects, HPLA was identical was identical to the α-peak component and MeHPLA was identical to the β-peak component.

## ROLE OF HPLA AND MHPLA IN CELL PROLIFERATION

The identification of HPLA and MeHPLA as ligands for nuclear type II binding sites has important implications with respect to potential cell growth regulation by these compounds. A number of studies from our laboratory have shown that the stimulation of nuclear type II binding sites in the rat uterus by estrogenic hormones is highly correlated with cellular hypertrophy, hyperplasia, and DNA synthesis (Markaverich and Clark, 1981; Markaverich et al., 1981, Clark et al., 1982). Therefore, we have suggested that nuclear type II binding sites may be involved in the modulation and/or regulation of cell growth. Further, we proposed the ligand for nuclear type II sites may also be involved in cell growth regulation through a direct inter- action with this binding site. Since studies in our laboratory have demonstrated that estradiol is unlikely to bind to nuclear type II sites in vivo because of its low binding affinity ($K_d \sim 20$ nmol) for this protein, we have suggested the function of nuclear type II site is to bind the type II ligand rather than estradiol (Markaverich et al., 1981, 1983, 1984; Markaverich and Clark 1979, Clark et al., 1982).

On the basis of the data presented in this study with authentic HPLA and MeHPLA, it can be concluded that MeHPLA is the more important of these two ligands with respect to interaction with type II binding sites in vivo. Competition analysis revealed that MeHPLA interacts with nuclear type II sites with a 30- to 40-fold higherf affinity than HPLA and the estimated $K_d$ for the MeHPLA for the MeHPLA-type II binding interaction is approximately 5 nmolar (calculated from data presented in Figure 1). However, it must be noted that this binding affinity is only approximate since it is estimated on the basis of [³H]estradiol binding data. Direct assessment of the binding of [³H]MeHPLA to nuclear type II sites must be performed to obtain a more accurate value. Nevertheless, it is apparent from the data presented in Figure 1 that MeHPLA interacts with nuclear type II sites with a much greater binding affinity than HPLA or estradiol. Therefore, one would predict that MeHPLA may be the more significant compound with respect to potential effects on cell growth regulation. This appears to be the case. When we assessed the effects of the authentic compounds on MCF-7 human breast cancer cell growth, MeHPLA (but not HPLA) was capable of inhibiting cell proliferation at the concentrations tested (Figure 2) and this effect was reversible following MeHPLA removal from the medium. Although the doses required for cell growth inhibition (1-10 μg/mL) appeared to be somewhate high, cell uptake studies with [³H]MeHPLA revealed that only 1% of the compound was taken up by MCF-7 cells under these experimental conditions. Therefore the calculated "intracellular concentrations" of MeHPLA in this experiment were approximately 10-100 ng/mL, which is in excellent agreement with the nuclear type II binding data (Figure 1). On the basis of these observations, it is evident that the MeHPLA inhibition of MCF-7 cell growth probably occurs through an interaction with nuclear type II sites, although the mechanisms remains to be determined. Failure of HPLA to inhibit MCF-7 growth (Figure 2) is consistent with our observation that this compound interacts with nuclear type II sites with a 30- to 40-fold lower affinity than MeHPLA.

## HPLA and Me-HPLA Competition for Nuclear Type II Sites

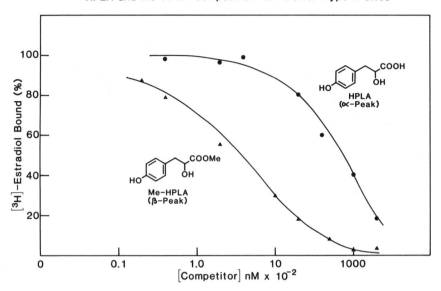

Figure 1. Competition Analysis with Authentic HPLA and MeHPLA for [³H]Estradiol Binding to Nuclear Type II Sites.

## Type II Ligand Effects on MCF-7 Cell Growth

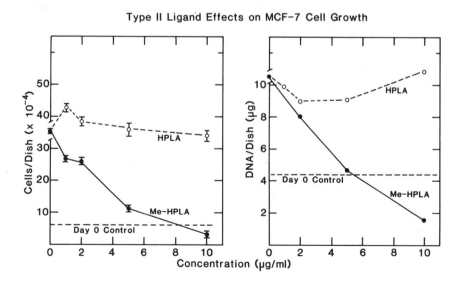

Figure 2. Effects of Authentic HPLA and MeHPLA on MCF-7 Human Breast Cancer Cell Growth In Vitro. Cell growth was evaluated by measuring the number of cells per dish and the total DNA content per dish.

Similarly, HPLA was not as effect as MeHPLA in inhibiting estradiol stimulation of uterine growth in the immature rat. The data in Figure 3 demonstrate that 10 μg of MeHPLA almost completely blocked estradiol stimulation of true uterine growth (assessed by comparing both wet and dry weight), whereas no significant response was obtained with 10 μg of HPLA. The antagonism observed with 100 μg of HPLA (Figure 3) is consistent with our binding competition data, suggesting that HPLA will interact with nuclear type II sites at high concentrations. We suspect that complete antagonism of estrogenic response would have been observed with higher doses of HPLA. Since 10μg of MeHPLA blocked uterine growth (Figure 3) and has a 30- to 40-fold greater affinity for the type II site than HPLA, one would predict that approximately 300-400 μg of HPLA would be required to exert the same effect as MeHPLA on uterine growth. It is also possible that the inhibition activity obtained for HPLA in vivo is actually due to its partial conversion to MeHPLA. This relationship is currently under detailed study.

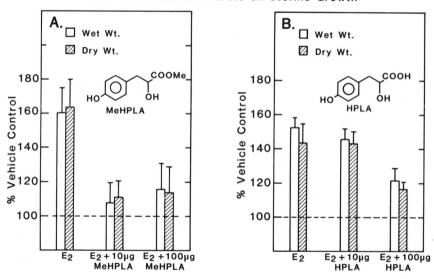

Figure 3. Effects of HPLA and MeHPLA on Estrogen-Stimulated Uterine Growth. Immature (21 day old) Sprague-Dawley rats (6-8 per group) were injected with saline-ethanol (20%) vehicle containing estradiol (0.5μg) with or without 10 or 100 μg of HPLA or MeHPLA. Animals were sacrificed 24 hours following injection and uterine wet and dry weights were determined.

Our observation that HPLA and MeHPLA may function to regulate cell growth through an interaction with nuclear type II sites has important implications with respect to regulation of normal and malignant cell growth. At present, we do not know the precise biological origin of HPLA and MeHPLA in rat tissues since HPLA can be formed by metabolism of both bioflavonoids (Griffiths and Smith, 1972) and tyrosine (Karoum 1985; Haley and Harper, 1982; Fuchs-Mettler et al., 1980). The relative contributions of these two pathways to the endogenous HPLA pool has not been clearly delineated. Likewise, methylation of HPLA to form MeHPLA has not been demonstrated, although MeHPLA has been isolated from human brain tissue (Lesch, 1979) and we have found the β-peak material in a variety of rat tissues. Interestingly, we have shown that MeHPLA is deficient in tumor cytosol preparations (Markaverich et al., 1984) whch contain HPLA. These data suggest that tumors are capable of metabolizing MeHPLA. We suspect that the deficiency of MeHPLA in rat, mouse, and human mammary tumors may result from inactivation via conversion to HPLA. A variety of tumor preparations (Schatz and Hochbert, 1981, Abul-Hajj and Nurieddin, 1983) and MCF-7 human breast cancer cells (Adams et al., 1986) have been shown to contain significant levels of esterase activity which may be responsible for this conversion.

Likewise, it is also possible that MeHPLA activity in the rat uterus may be under estrogenic regulation. Hochberg and coworkers have demonstrated that rat, rabbit, and bovine uteri contain significant levels of esterase activity which is likely to be responsible for lipoidal estrogen synthesis (Hochberg et al., 1977, 1979; Mellon-Nassbaum and Hochberg, 1980). It is tempting to speculate that uterine esterases are under estrogen regulation and may be responsible for converting MeHPLA to HPLA, the form of the type II ligand with lower biological activity. Therefore, it is possible that estrogens modulate uterotropic response via regulation of the intracellular MeHPLA pools. A decrease in the concentration of MeHPLA in estrogen target cells is likely to lead to the expression of nuclear type II site function which is directly correlated with estrogenic stimulation of cellular hypertrophy, hyperplasia, and DNA synthesis. If this model is correct, one would predict that the deficiency of MeHPLA in malignant cells may also be directly related to the uncontrolled rate of proliferation in these cell populations. Whether or not this deficiency results from high levels of tumor esterase activity, or from defects in HPLA synthesis remains to be resolved. Our current data support the former hypothesis since we are unable to detect a deficiency of HPLA in tumor cell cytosol preparation. Therefore, the uncontrolled proliferation of malignant cells is directly related not only to a permanent stimulation of nuclear type II sites (Watson and Clark, 1980; Watson et al., 1980; Syne et al., 1982a&b), but also to very low to nonmeasurable levels of MeHPLA. Further, the high levels of type II sites measured in tumor nuclei may be due to a deficiency of MeHPLA, such that more free binding sites are available to be readily detected by [$^3$H]estradiol exchange.

We are particulary intrigued by a possible relationship between dietary precursors of HPLA and the growth and proliferation of tumor cells that might be mediated by MeHPLA. There are many scattered reports in the literature on the antitumor activity of anthocyanins and other bioflavoinoid (Markaverich et al., 1988 a&b; Wattenberg, 1983; Carr, 1985; Picardo et al., 1987). The antitumor activity reported has generally been weak, and critics have pointed out that bioflavonoids are not absorbed intact from the diet. Indeed, the antitumor activity attributed to the bioflavonoids is must greater _in vitro_ than _in vivo_. The general consensus has been that the data for antitumor activity _in vivo_ are less than compelling. Our research sheds new light on this dilemma. HPLA is a known metabolite of several bioflavonoids, so it is likely that the antitumor activity of the bioflavonoids is actually due to MeHPLA. There have also been several reports on the antitumor activity of catechols (Wallenberg 1983; Carr, 1985, Picardo et al., 1987). Again, the evidence is weak. However, it is possible that MeHPLA is one of the more potent endogeneous members of this family of antitumor agents. Since MeHPLA is an endogenous metabolite and probably has few, if any, side-effects it has not escaped our attention that MeHPLA and related compounds may be of value as prophylatic agents for protection against cancer.

## ACKNOWLEDGEMENTS

This work was supported by grants from the National Institutes of Health (CA-35480,HD-08436), the American Institute for Cancer Research (86 B14), and The American Cancer Society RD-266.

**REFERENCES**

Abul-Hajj, Y.J. and Nurieddin, A. Steroids 42:417 (1983).

Adams, J.B., Hall, R.T., and Nott, S. J. Steroid Biochem. 24:1159 (1936).

Carr, B.I. Cancer (Suppl.) 55:218 (1985).

Clark, J.H., Williams, M., Upchurch, S., Eriksson, H., Helton, E., and Markaverich, B.M. J. Steroid Biochem. 16:323 (1982).

Fuchs-Mettler, M., Curtins, H.-C., Baerlocher, K., and Ettlinger, L. Europ. J. Biochem. 108:527 (1980).

Griffiths, L.A. and Smith, G.E. Biochem. J. 128:901 (1972).

Grzycka, K. and Milkowska, J. Ann. Univ. Mariae Curie-Sklodowska, Sect. D 32:339 (1977)

Haley, C.J. and Harper, A.E. Metabolism 31:524 (1982).

Hochberg, R.B., Bandy, L., Ponticorvo, L., and Lieberman, S. Proc. Natl. Acad. Sci. 74:941 (1977).

Hochberg, R.B., Bandy, L., Ponticorvo, L., Welch, M., and Lieberman, S. J. Steroid Biochem. 11:1333 (1979).

Karoum, F. Biogenic Amines 2:d269 (1985).

Lesch, P. Deut. Med. Wochenschr. 98:1929 (1979).

Markaverich, B.M. and Clark, J.H. Endocrinology 105:14582 (1979).

Markaverich, B.M., Gregory, R.R., Alejandro, M.A., Clark, J.H. Johnson, G., and Middleditch, B.S. J. Biol. Chem. 263:7203 (1988a).

Markaverich, B.M., Roberts, R.R., Alejandro M.A., Johnson, G.A., Middleton, B.S. and Clark, J.H. J. Steroid Biochem. 30:1 (1988b).

Markaverich, B.M., Roberts, R.R., Alejandro, M.A., and Clark, J.H. (1984) Cancer Res. 44:1575.

Markaverich, B.M., Roberts, R.R., Finney, R.W., and Clark, J.H. J. Biol. Chem. 258:11663 (1983).

Markaverich, B.M., Upchurch, S., and Clark, J.H. J. Steroid Biochem. 14:125 (1981).

Mellon-Nussbaum, S. and Hochberg, R.B. J. Biol. Chem. 255:5566 (1980).

Molnar, J., Foldeak, S., Domonkos,S., Schneider, B., Forczek, E. and Beladi, I. Srjtosztodas Farmakol. 8:55 (1979).

Picardo, M., Passi, S., Nazzaro-Porro, M., Breathnach, A., Zompetta, C., Faggioni, A. and Riley, P. Biochem. Pharmacol. 36:417 (1976).

Schatz, F. and Hochberg, R.B. Endocrinology 109:697 (1981).

Syne, J.S., Markaverich, B.M., Clark, J.H., and Panko, W.B. Cancer Res. 42:4443 (1982a).

Syne, J.S., Markaverich, B.M., Clark, J.H., and Panko, W.B. Cancer Res. 42:4449 (1982b).

Watson, C.S. and Clark, J.H. J. Receptor Res. 1:91 (1980).

Watson, C.S., Medina, D., and Clark, J.H. Endocrinology 107:1432 (1980).

Wattenberg, L.W. Cancer Res. (Suppl.) 43:2448s (1983).

Wattenberg, L.W. and Leong, J.L. Cancer Res. 30:1922 (1970).

PROMOTER SPECIFIC ACTIVATING DOMAINS OF THE CHICKEN PROGESTERONE
RECEPTOR

Orla M. Conneely, Denise Kettelberger, Ming-Jer Tsai and Bert W.
O'Malley, Department of Cell Biology, Baylor College of Medicine,
Houston, Texas 77030

Summary

Analysis of the functional properties of two naturally occurring
chicken progesterone receptors (cPR) has allowed us to begin to
examine the interaction of steroid receptors with transcription
factors. The cPR A and B proteins differ by an additional 128 amino
acids located at the N-terminus of the B protein. These proteins are
functionally distinct in terms of their ability to activate specific
traget genes. The N-terminal region of the cPR B protein is an acidic
domain which acts as a promoter specific transcriptional regulating
domain when transferred to a heterologous receptor. The regulatory
function of this domain is independent of the DNA binding specificity
of the receptor and appears to reflect modulation of receptor
interaction with transcription factors. The estrogen and progesterone
receptors appear to interact with common transcription factors.
Competition for common transcription factors by these proteins does
not require DNA binding.

Introduction

Steroids regulate gene expression in eucaryotic cells by binding to
specific intracellular receptors which, in turn, modulate the
transcription of target genes (Eriksson and Gustaffson, 1983; Schrader
et al, 1981; Yamamoto, 1985). The steroid receptors are part of a
superfamily of ligand inducible transcription factors (Evans, 1988).
The proteins are all structurally related consisting of a steroid
binding domain and a DNA binding domain located within a 300 amino
acid sequence extending from the carboxy terminal end of the
proteins. The least conserved region of the receptors is located on
the N-terminal side of the DNA binding domain. This region is
hypervariable between receptors in terms of length and amino acid
composition, and corresponds to the immunogenic domain of these
proteins. Several lines of evidence indicate that this region is
involved in modulating transcriptional activity of different target
genes by the steroid receptors (Giguere et al, 1986; Rusconi et al,
1987; Kumar et al, 1987; Tora et al, 1988).

The chicken progesterone receptor is unusual in that it consists of two hormone binding proteins which arise from the same gene (Huckaby et al, 1987) and occur in roughly equivalent amounts in chick oviduct (Schrader et al, 1972). The region which differs the A from the B proteins is a 128 amino acid sequence located at the N-terminus of the B protein. The sequence has two in frame AUG signals in this region; the first is used to translate the B protein, and the second is located immediately downstream from the B-specific N-terminal region (Conneely et al, 1987a).

In the present study we demonstrate, using site-directed mutagenesis, that the receptor A protein is produced by alternate initiation of translation at the internal AUG signal located at methionine 129 in the amino acid sequence of the receptor B protein. In addition, we show that the A and B proteins are functionally distinct in terms of their ability to preferentially induce transcription of specific target genes. The N-terminal 128 amino acid region which is specific to the B protein consists of an acidic domain which is capable of acting as a promoter specific activator domain. Furthermore, this regulation appears to be independent of specificity of DNA binding by the receptor since transfer of this acidic domain to the estrogen receptor alters promoter activity in response to this receptor.

### The cPR A and B Proteins Arise by Alternate Initiation of Translation of a Single mRNA

The full-length cPR cDNA encodes the receptor B protein which is 787 amino acids in length (Conneely et al, 1987a). The region which differentiates in B and A proteins is a 128 amino acid sequence at the N-terminus of the B protein. The two in-frame AUG signals in this region are both encoded by the first exon of the chromosomal gene (Huckaby et al, 1987). A monoclonal antibody which recognizes only the B protein reacts with an epitope encoded by sequence within 90 nucleotides upstream from the internal AUG signal (Conneely et al, 1987a). These data taken together indicate that alternate splicing is an unlikely mechanism to generate the A and B proteins since the N-terminal of both proteins is contained within a single exon. The receptor A protein may therefore arise by proteolysis of the larger B protein or by alternate initiation of translation at the internal AUG codon.

Our initial suggestions that the A and B proteins appear to arise by alternate initiation of translation came from the following evidence: Expression of the full-length cPR cDNA in heterologous cells or in a cell free system gives rise to both the A and B proteins (Conneely et al, 1987b). Furthermore, deletion of the 5' end of the cDNA to generate truncated mRNAs which lack the initiation signal for translation of the receptor B protein results in expression of a functional receptor A protein. In order to confirm that the internal AUG signal provided the translational start site for the A protein, we used oligo-nucleotide directed mutagenesis to introduce a single amino acid mutation into the cPR cDNA. The sequence of the mutated region

is shown in figure 1. The ATG triplet which encodes methionine 129
was replaced by a GTC triplet resulting in a single conservative amino
acid substitution from methionine to valine at position 129 in the
receptor sequence. The effect of this mutation on the ability of the
mRNA to produce the cPR A and B proteins then was examined by
expression of the cDNA in heterologous receptor negative cells.

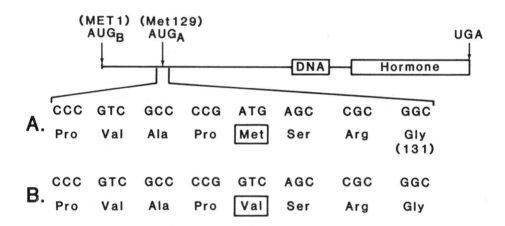

Figure 1. Schematic representation of the cPR cDNA and comparison of
wild-type and mutated sequences in the region surrounding AUG$_A$.
A synthetic oligonucleotide (sequence B) encoding amino acids 125-131
in the receptor cDNA was used to direct a site-specific two-nucleotide
substitution at the ATG triplet encoding methionine 129 in the cPR
cDNA. The cDNA template for mutagenesis was prepared by subcloning a
5' PstI – SstI fragment of the receptor cDNA into M13mp18. Mutagenesis was carried out according to the method of Kunkel *et al*,
(1987) using the 24-mer oligonucleotide sequence B as primer. Mutants
were detected by screening with the mutant 24-mer which was 5' end
labeled with $^{32}$P and confirmed by dideoxy sequencing. Sequence
analysis showed that the oligonucleotide directed mutagenesis resulted
in a two base-pair change from ATG to GTC. The mutated fragment was
then used to replace the 5' region of the full-length cPR cDNA and the
receptor mutant (CPRβΔK) contained a single conservative amino acid
change from methionine to valine at position 129.

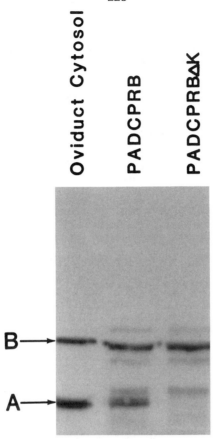

Figure 2.  Western immunoblot analysis of extracts from COS-M6 cells transfected with wild-type and mutant receptor cDNA constructs.
The complete cPR cDNA and mutant described in figure 1 were subcloned into the eucaryotic expression vector, p91023(B) (Wang *et al*, 1985) to generate the wild-type (pADCPRB) and mutant (pADCPRBΔK) receptor expresson constructs.  COS-M6 cells were transfected with 20 μg of either pADCPRB or pADCPRBΔK using the DEAE dextran method (Sampayrac *et al*, 1981).  48 hours after transfection, the cells were harvested and high salt extracts were prepared as previously described (Carson *et al*, 1987).  Western immunoblot analysis was carried out using mouse anti-cPR IgG, PR22.  Lane 1 shows the chicken cPR A and B proteins detected in cytosolic extracts from chick oviduct tissue as markers for this study.  Lanes 2 and 3 show the immunoreactive proteins detected in COS-M6 cells after transfection with the wild-type (pADCPRB, lane 2) and mutant receptor expression constructs (pADCPRBΔK, lane 3).

In these studies receptor negative Cos m-6 cells were transfected with expression plasmids containing either the wild type (pADcPRB) or mutant (pADcPRBΔK) cDNA. The receptor proteins expressed from the constructs were analyzed by Western immunoblot analysis of high-salt extracts from the transfected cells. The results of this analysis are shown in figure 2. The chick oviduct cPR A (72 KDa) and B (86 KDa) proteins detected in oviduct cytosol are shown in lane 1 for reference. The receptor proteins expressed from the wild-type and mutated cDNAs are shown in lanes 2 and 3, respectively. While both the A and B proteins were expressed from the wild-type cDNA, no receptor A protein was detected in extracts from cells transfected with the mutated cPR cDNA expression construct. Thus, a single conservative amino acid substitution from methionine to valine at position 129 in the amino acid sequence completely abolished translation of the receptor A protein. These results are consistent with the interpretation that the A protein arises by initiation of translation at the internal AUG signal encoding Methionine 129

## The cPR A and B Proteins are Functionally Distinct

The region which differentiates the cPR A and B proteins is the additional 128 amino acids located at the N-terminus of the receptor B protein. Both proteins bind progesterone with the same affinity and contact the same sites on a GRE/PRE oligonucleotide, a progesterone response element from the tyrosine amino transferase gene (Tsai *et al*, 1988). Several lines of evidence have suggested that the N-terminal region of the steroid receptors is involved in modulating transcriptional response to specific target genes (Giguere *et al*, 1986; Rusconi and Yamamoto, 1987; Kumar *et al*, 1987; Tora, *et al*, 1988). The single point mutation in the cPR cDNA sequence allowed us to selectively express the receptor B protein in heterologous receptor negative cells and to determine the functional activities of the A and B proteins using different target genes. For these studies, the receptor expression plasmids were cotransfected into CV-1 cells together with a reporter plasmid consisting of a progesterone response element and promoter fused to chloramphenicol acetyl transferase (CAT) as the reporter gene. Two target genes were tested; the PRE/GRE element from the tyrosine amino transferase gene fused to the thymidine kinase promoter in the plasmid PRETKCAT and a progesterone responsive promoter region containing 5' flanking sequences (-732/+40) of the chicken ovalbumin gene in the plasmid POVCATMA. The receptor expression plasmids were as follows: PADA, which contains a truncated cPR cDNA and expresses only the A protein, and pADcPRB and pADcPRBΔK which contain the wild type and mutated cPR cDNAs described above and express both B and A proteins and the B protein, respectively. The results of these functional analyses are shown in figure 3, panels A and B. The PRETKCAT reporter gene responded to both the A and B proteins when introduced separately into the cells (lanes 2 and 3) or together from the pADcPRB expression plasmid (lane 4). However, when the ovalbumin promoter was used as target gene a different pattern of induction was observed (panel B). While this promoter did respond to the A protein (lane 2), no activity was observed when the receptor

# A. PRETKCAT

# B. POCATMA

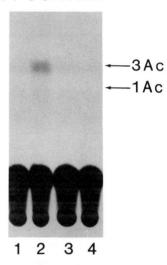

3Ac→

1Ac→

←3Ac

←1Ac

1 2 3 4

1 2 3 4

Figure 3.  Differential induction of steroid-regulated promoters by progesterone receptors A and B.

CV-1 cells were co-transfected with a reporter plasmid and a receptor expression plasmid in the presence of $10^{-7}$M progesterone using 5 µg/ml polybrene (Chaney et al, 1986), followed by a 25% glycerol shock. Cells were harvested 48 hours after transfection and assayed for chloramphenicol acetyl transferase (CAT) activity using standard protocols (Gorman et al, 1982) Panel A:  The reporter plasmid was PRETKCAT (Jantzen et al, 1987) which contained two copies of the PRE-GRE sequence of the tyrosine amino transferase (TAT) gene in reverse orientation fused to the thymidine kinase promoter (TK) and chloramphenicol acetyl transferase gene (CAT) in the vector PBLCAT2. This plasmid (5 µg) was co-transfected with receptor expression vectors shown in lanes 1-4 (1, vector without cDNA insert; 2, PADCPRA; 3, PADCPRB; 4, PADCPRBΔK).  Panel B:  The reporter plasmid POVCATMA (10 µg) which contained -732 to +41 of the ovalbumin gene fused to CAT was co-transfected with 5 µg of each receptor expression plasmid as shown for panel A.

B or the wild type cDNA expression constructs were used in the transfections (lanes 3 and 4). No activity was observed with either target gene when cotransfected with plasmid vector without receptor cDNA (panels A and B, lane 1). These data demonstrated a differential effect of the A and B proteins with regard to specificity of target gene activation. Furthermore, since transcription of the ovalbumin promoter was not activated when both the A and B proteins were expressed together, the B protein also may compete out the activation of this promoter by the A protein. In order to investigate this competition further we examined the response of the ovalbumin promoter to increasing concentrations of the wild type receptor expression plasmid (figure 4, panels A-C). While the promoter response to the A protein increased rapidly to a maximum level with increasing concentrations of plasmid (panel A), no CAT activity was detected at any concentration of the wild-type pADcPRB expression plasmid (panel B) or with the control vector plasmid without cDNA insert (panel C). It appears, therefore, that the B protein can inhibit activation of the ovalbumin promoter by the A protein.

## Ovalbumin CAT

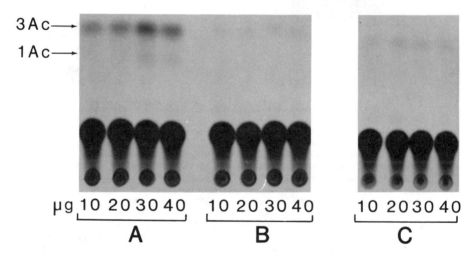

Figure 4. Progesterone receptor B inhibition of ovalbumin transcriptional response to receptor A.
The reporter plasmid POVCATMA was co-transfected with increasing doses of receptor expresson vectors (1, PADCPRA (A protein); 2, PADCPRB (A and B proteins); 3, P91023(B) vector without cDNA insert).

The differential activities of the A and B proteins observed above indicate that the 128 amino acid N-terminal region of the receptor plays a role in determining the specificity of target gene activation by the progesterone receptor. Since the A and B proteins are two naturally occurring isoforms of the progesterone receptor and are found in roughly equal concentrations in progesterone responsive cells, their relative inductive capacity for specific target genes is likely to be of physiological significance.

## The N-terminal Acidic Domain of the Progesterone Receptor is a Promoter Specific Transcriptional Regulator Domain

In order to further examine the regulatory role of the N-terminal acidic domain of cPR, we decided to transfer this domain to the N-terminal end of the human estrogen receptor and to determine its effect on specificity of target gene activation by a heterologous receptor which interacted with different response elements. In these experiments, we fused a 5' fragment of the cPR cDNA encoding the N-terminal 128 amino acids to the 5' end of the human estrogen receptor (hER) to generate the chimeric receptor expression construct PRBER. We then compared the functional activity of this chimera to that of hER using two different target genes. The target plasmids used for transfections are essentially the same as those described above. The first VERETKCAT, is similar to PRETKCAT but included a DNA fragment from the vitellogenin gene containing estrogen response elements to replace the PRE/GRE in PRETKCAT. The second target is POVCATMA used in the previous experiments and which also contains an estrogen response element. Each of these target plasmids was cotransfected into CV-1 cells with either the hER expression plasmid ($\Delta$HER) or the chimeric receptor expression plasmid (PRBER). Transcriptional response was measured by the CAT activity observed in each case. The results are shown in figure 5, panels A and B. Comparison of the VERETKCAT responses to human estrogen receptor and the PRBER chimera (panel A, lanes 1 and 2) showed a higher CAT activity expressed by the chimera indicating that the acidic domain of the progesterone receptor potentiates the response of this target gene to the estrogen receptor. The results of similar analysis using the ovalbumin promoter (POVCATMA) as target element are shown in panel B. In this case, while the ovalbumin promoter does respond to the estrogen receptor (lane 2), no activity was detected from this promoter when the chimera was used as a transactivator (lane 3). These results indicate that the acidic domain of cPR abolishes the transcriptional response of the ovalbumin promoter to the estrogen receptor. Thus, the acidic domain acts as a promoter specific activator/inhibitor domain when transferred to a heterologous receptor. Since the estrogen and progesterone receptors bind the target DNA through different hormone response elements, this regulatory effect of the acidic domain appears to be independent of specificity of DNA binding by the receptors. These observations also suggested that the inhibitory effect of the B protein on ovalbumin activation by the cPR A receptor may also be independent of DNA binding.

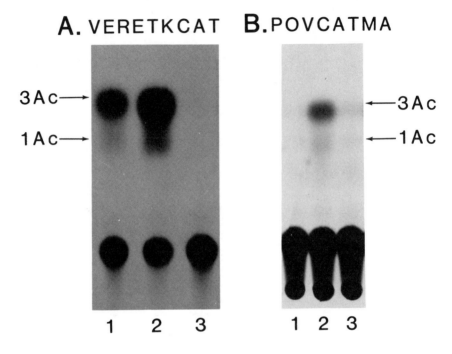

# A. VERETKCAT   B.POVCATMA

Figure 5. The N-terminal acidic domain of the progesterone receptor B protein is a promoter specific transcriptional regulating domain.
The chimeric construct PADPRBER was generated by excising a 5' PVUI-NAEI fragment from the cPR cDNA mutant (CPRβΔK, see figure 1). This fragment was blunt-end ligated to the human ER cDNA at a TTHIIII site located 30 nucleotides upstream of the initiator AUG such that the complete ER open reading frame was maintained. Thus fusion gave a chimeric receptor consisting of the acidic 128 N-terminal amino acids of the cPRB protein linked to the complete hER protein. The functional specificity of this chimera was compared to that of the human estrogen receptor (hER) by co-transfecting either the chimera (PADPRBER) or hER expression plasmids (PADΔHER) with each of two target plasmids (10 μg). The target genes were VERETKCAT which contained the estrogen response element from the vitellogenin gene (-331/-297) fused to the TK promoter described above (see legend figure 3) and the same POVCATMA ovalbumin target plasmid used in the previous experiments.

The Estrogen and Progesterone Receptors Interact with Common Transcription Factors

The analyses described above indicated that regions within the N-terminal region of the receptors may interact with transcription factors to alter the rate of transcription of specific target genes and that this interaction is independent of the specificity of DNA binding by the receptors. Two of the target genes we had analyzed, VERETKCAT and PRETKCAT differed only in the hormone response elements which were fused to the thymidine kinase promoter. The response of these genes to estrogen or progesterone receptors is then determined by the specific steroid response element fused to the promoter. However, since the N-terminal regions of these receptors are not conserved, we decided to examine whether these receptors interacted common transcription factors to activate the thymidine kinase promoter. The response of PRETKCAT to increasing concentrations of progesterone receptor B expression plasmid (pADCPRβΔK) is shown in figure 6. The transcriptional response to the receptor increases initially in a dose dependent fashion and then plateaus at higher receptor plasmid concentrations while the receptor concentration continues to increase. These data suggest that additional factors required for transcription may be saturated at these receptor levels. If increasing concentrations of estrogen receptor expression plasmid (ΔHER) are now cotransfected with the progesterone receptor, the response of PRETKCAT to the progesterone receptor is dramatically decreased to a minimum basal level, even though the estrogen receptor does not interact with this PRE element or activate PRETKCAT. These data indicate that in the case of this promoter, the estrogen and progesterone receptors appear to interact with common transcription factors. Furthermore, this interaction is independent of DNA binding since the estrogen receptor does not bind to the progesterone response element on PRETKCAT. The estrogen receptor inhibits the activation of a progesterone response element by its receptor in a dose dependent fashion. We have yet to establish whether this competitive effect is promoter specific or if this effect is hormone dependent.

A similar phenomenon referred to as a "squelch effect" previously has been observed in the case of the yeast transcriptional activator Gal 4 (Gill and Ptashne, 1988). In these studies, strong activating regions of Gal 4 inhibit transcription of certain genes which lack Gal 4 binding sites. It is interesting to note that these activating regions are reminiscent of the acidic domain of cPRB. The squelch effect of the Gal 4 activating regions is independent of DNA binding and appears to reflect titration of a transcription factor. This type of phenomenon may therefore reflect a general mechanism of regulation by the steroid receptors and yeast transcriptional activators alike. In any case, this type of system should provide us with a means by which we can begin to look more directly at how steroid receptors interact with transcription factors.

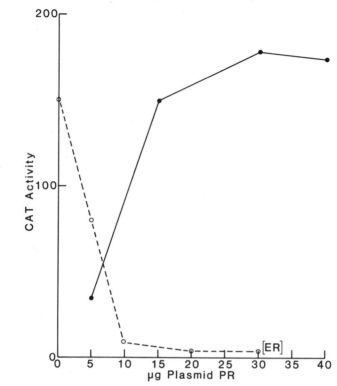

Figure 6. The estrogen and progesterone receptors interact with common transcription factors.
10µg PRETKCAT was co-transfected into CV-1 cells with increasing concentrations of progesterone receptor expression plasmid (PADCPRβΔK). CAT activity was measured by counting the [14]C acetylated products of chloramphenicol recovered from TLC plates. The results are shown by the solid line in the figure. The effect of hER on this response was measured by co-transfecting increasing concentrations of PADΔHER with 10 µg of PADCPRβΔK and 10 µg PRETKCAT reporter plasmid. The CAT activity detected is shown by the dotted line.

## References

Carson, M.A., Tasi, M.-J., Conneely, O.M., Maxwell, B.L., Clark, J.H., Dobson, A.D.W., Elbrecht, A., Toft, D.O., Schrader, W.T., and O'Malley, B.W. (1987) Mol. Endo. 1, 791-801.

Chaney, W.G., Howard, R.R., Pollard, J.W., Sallustio, S. and Stanley, P. (1986) Somatic Cell and Mol. Gen. 12, 237-244.

Conneely, O.M., Dobson, A.D.W., Tsai, M.-J., Beattie, W.G., Toft, D.O., Huckaby, C.S., Zarucki, T., Schrader, W.T. and O'Malley, B.W. (1987a) Mol. Endo.1 1, 517-525.

Conneely, O.M., Maxwell, B.L., Toft, D.O., Schrader, W.T., and O'Malley, B.W. (1987b) Biochem. Biophys. Res. Commun. 149, 493-501.

Eriksson, H. and Gustafsson, J.A. Eds. (1983) Steroid Hormone Receptors: Structure and Function, Novel Symposium 57 (Elsevier, Amsterdam).

Evans, R.M. (1988) Science 240, 889-895.

Giguere, V., Hollenberg, S.M., Rosenfeld, G.M., and Evans, R.M. (1986) Cell 46, 645-652.

Gill, G. and Ptashne, M. (1988) Nature 334, 721-724.

Gorman, C.M., Moffat, L.F., and Howard, B.H. (1982) Mol. Cell Biol. 2, 1044-1051.

Huckaby, C.S., Conneely, O.M., Beattie, W.G., Dobson, A.D.W., Tsai, M.-J. and O'Malley, B.W. (1987) Proc. Natl. Acad. Sci 84, 8380-8384.

Jantzen, H.N., Strähle, U., Gloss, G., Stewart, F., Schmid, W., Boshart, M. Miksikek, R. and Schütz, G. (1987) Cell 49, 29-38.

Kumar, V., Green, S., Stack, G., Berry, M., Jin, J.R. and Chambon, P. (1987) Cell 51, 941-951.

Kunkel, T.A., Roberts, J.D. and LaKour, R.A. (1987) Methods Enzymol. 154, 367-382.

Rusconi, S. and Yamamoto, K.R. (1987) EMBO J. 6, 1309-1315.

Sampayrac, L.M. and Danna, K.J. (1981) Proc. Natl. Acad. Sci. 78, 7575-7578.

Schrader, W.T., Birnbaumer, M.E., Hughes, M.R., Weigel, N.L., Grody, W.W. and O'Malley, B.W. (1981) Recent Prog. Horm. Res. 37, 583-633.

Schrader, W.T. and O'Malley, B.V. (1972) J. Biol. Chem. 247, 51-59.

Tora, L., Gronemeyer, H., Turcotte, B., Gaub, M.-P. and Chambon, P. (1988) Nature 333, 195-188.

Tsai, S.Y., Carlstedt-Duke, J., Weigel, N.L., Dahlman, K., Gustafsson, J.A., Tsai, M.-J. and O'Malley, B.W. (1988) Cell 55, 361-369.

Wang, G.G. et al (1985) Science 228, 810-815.

Yamamoto, K.R. (1985) Ann. Rev. Genetics 19, 209-252.

## Discussion

Tata:      In your ovalbumin-CAT construct transfection experiments, have you considered the possibility that the B form of CPR recognizes a negative regulatory sequence that the A form does not?

Conneely:  This is indeed a possible explanation for the differential ovalbumin response to the two proteins. Further deletion analysis of the ovalbumin promoter will allow us to answer this question.

Schmid:    Was the construct used for the competition of PR by ER the TK promoter?

Conneely:  The promoter was the TK promoter; we have not tried other promoters as yet.

Roy:       I have a question concerning the estrogenic inhibition of the progesterone response in the transfected cells. You are suggesting that there may be some sort of competition between transacting factors which are needed for the action of both of these two hormone receptors. It is not clear to me whether you are thinking of transcription factors which are needed by all genes or some sort of specific factors common to only a small group of genes.

Conneely:  The TK promoter used in these studies contained two SP1 binding sites, a CCAAT box and a TATA box. The competition between the receptors is therefore likely to be for transcription factors required to bind one or more of these elements.

Thompson:  In the immunoblots of $PR_A$ and $PR_B$ using wild-type and mutant genes to produce their respective proteins, what is the heavy band that lies between A and B?

Conneely:  The additional minor bands observed above the A and B proteins may arise from weak non-AUG initiation sites. These bands are distinguished from the A protein by their reactivity with an anti-CPR monoclonal IgG specific for the B protein.

Thompson:  In ER/PR competitive experiments, did you vary the PR concentration and hold ER constant?

Conneely:  No.

Simons:    In the rabbit, there does not appear to be alternative start sites so that only the B receptor is formed. How general do you think alternative start sites are for the progesterone receptor and to what purpose do you think these alternative start sites serve?

Conneely:   The rabbit and human progesterone receptor sequences are
            not conserved with cPR in this region. The cPR contains a
            19 glutamic acid residue repeat in the B specific N-
            terminal region. This acidic structure is reminiscent of
            yeast activating domains and acts as a promoter specific
            transcriptional regulator. One could therefore speculate
            that the use of alternate start sites to produce two
            proteins would have a physiological role in differential
            regulation of target genes by two proteins with different
            target gene specificities.

Simons:     So far, there has not been found any antiprogestins for the
            chicken. Have you looked at whether classical
            antiprogestins work with the chick receptor in the
            transient transfection assays and, if they are active, is
            there any difference in the interactions of the
            antiprogestins with the A and B forms of the receptor.

Conneely:   We have not tested any antiprogestins in this system yet.

# Index